Atlas of Design

Vol. 5 »

editors
Brooke E. Marston
Caroline Rose
Nat Case
Vanessa Knoppke-Wetzel

published by
nacis
the North American
Cartographic Information Society

Atlas of Design, Volume Five

First Printing, October 2020

Editors: Brooke E. Marston, Caroline Rose, Nat Case, and Vanessa Knoppke-Wetzel

Published by the North American Cartographic Information Society, *nacis.org*

The type in this book has been set in Avenir, by Adrian Frutiger.

Photo of Leo Dillon by Hans van der Maarel.

Printed in the United States of America by Shapco Printing, Golden Valley, MN.

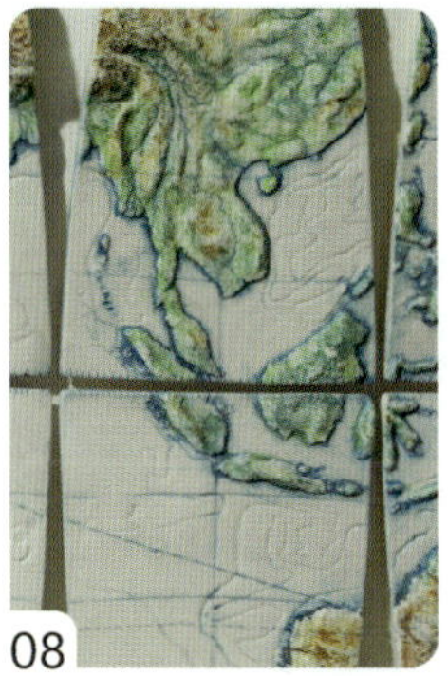
08

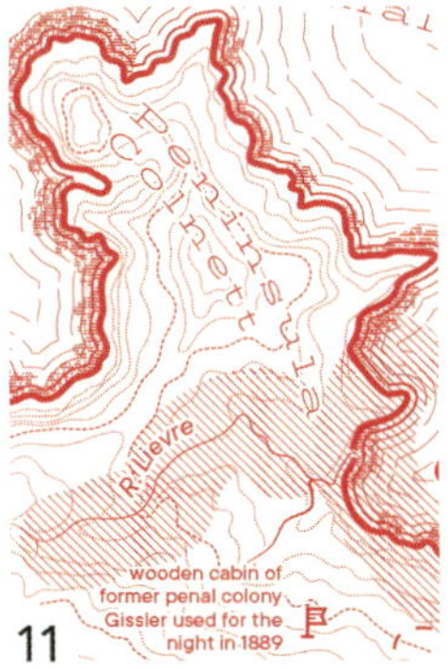

11

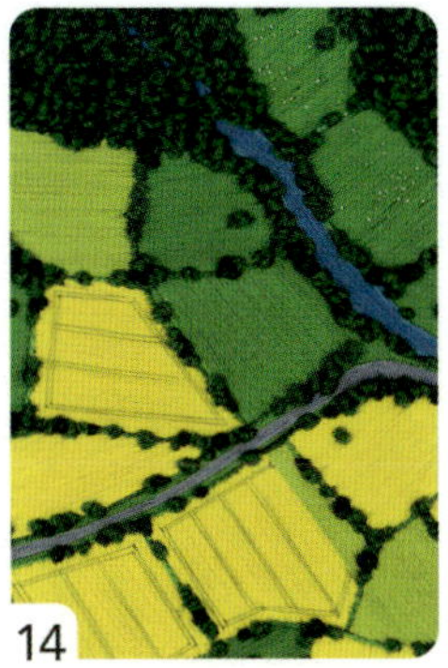
14

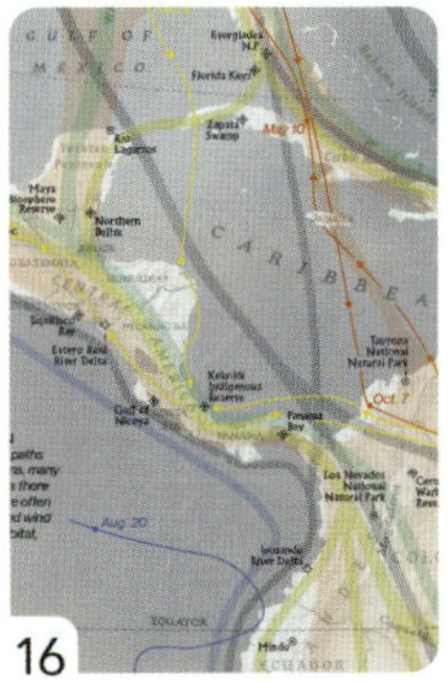
16

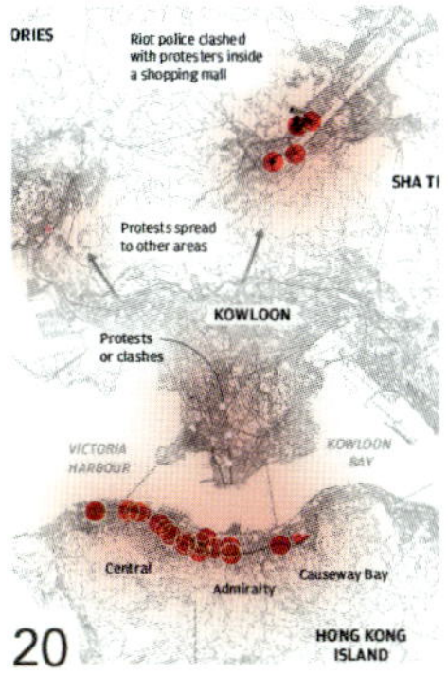

20

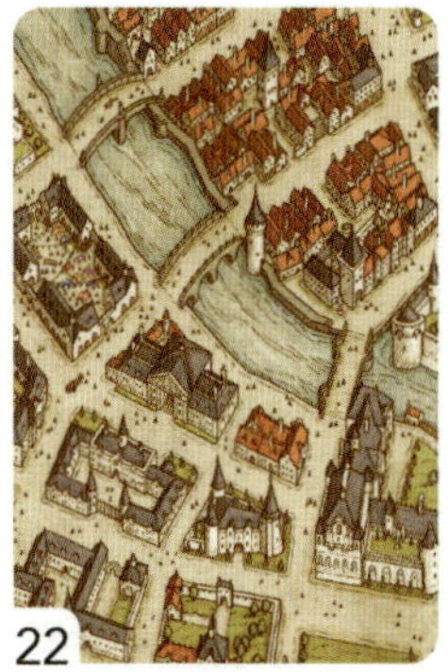
22

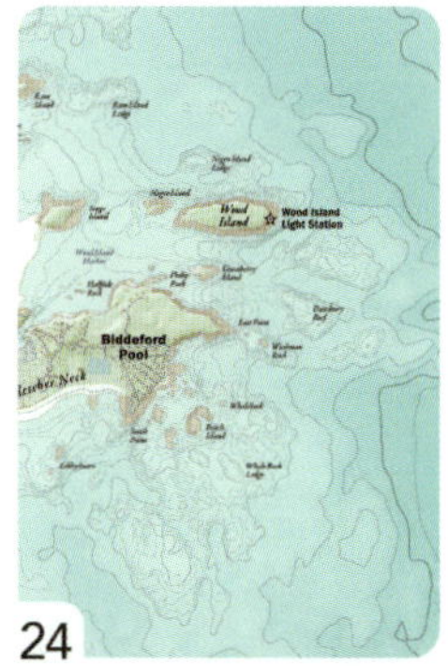
24

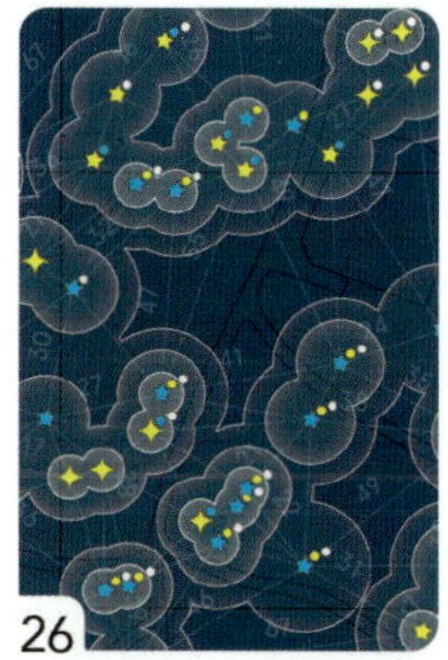
26

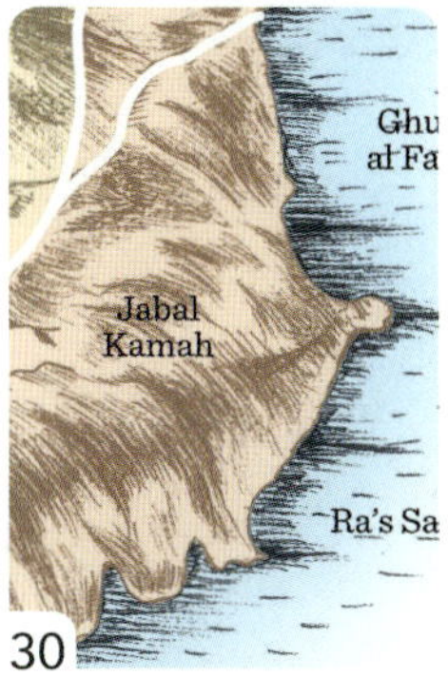

30

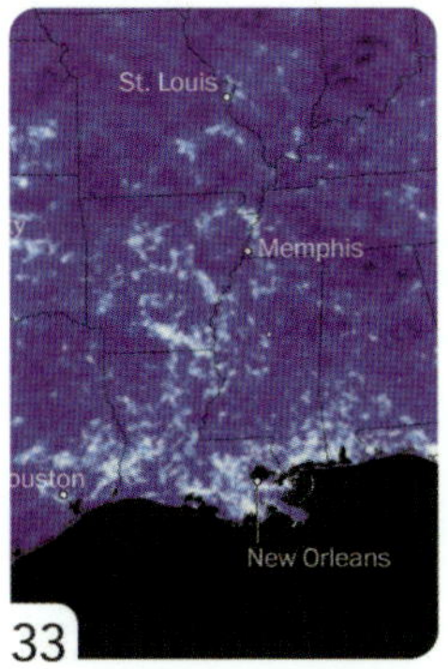

33

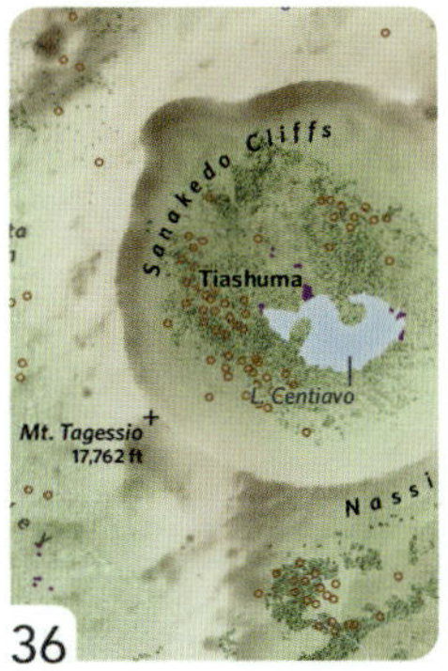

36

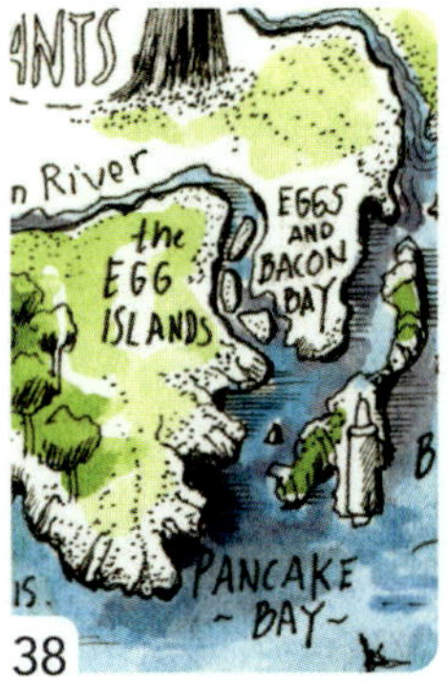

38

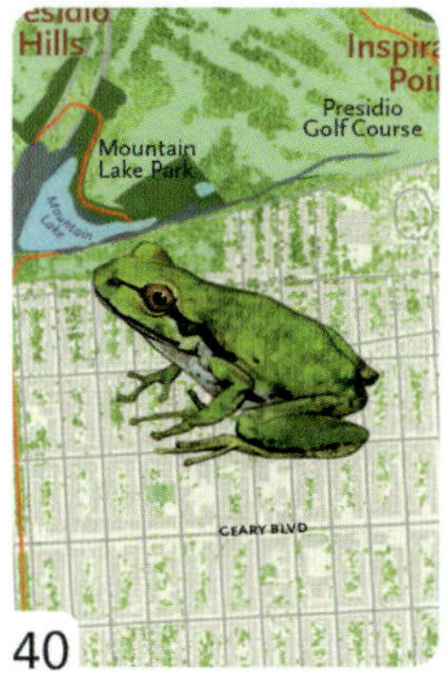

40

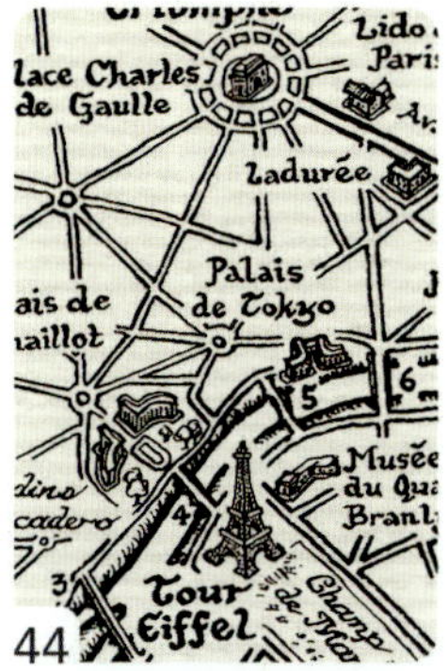

44

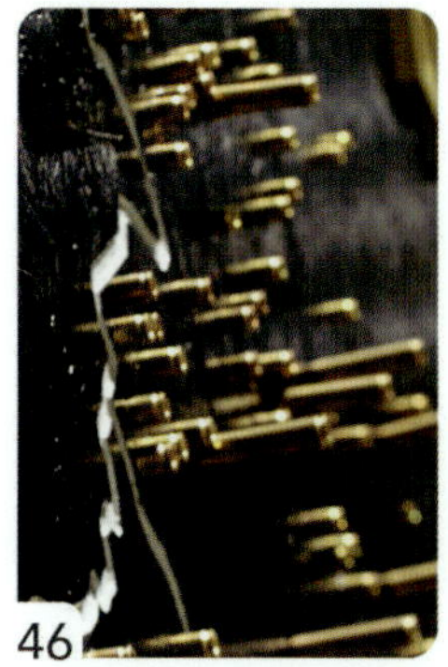
46

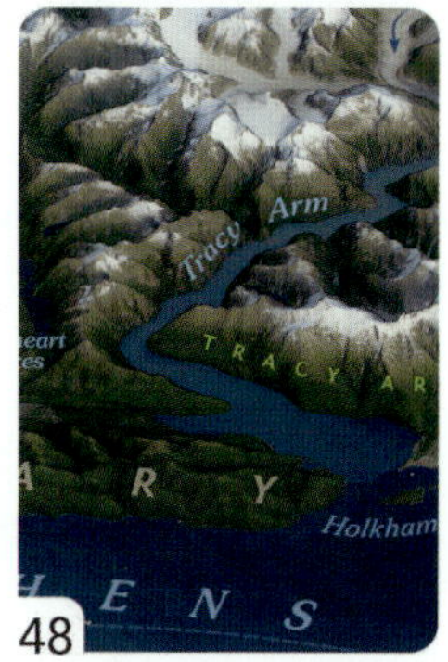

48

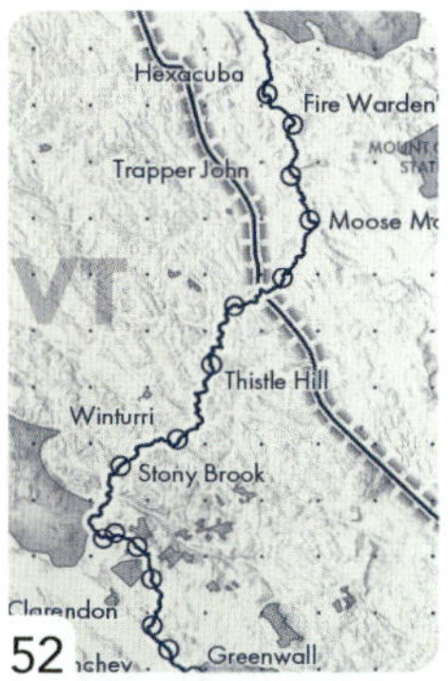

52

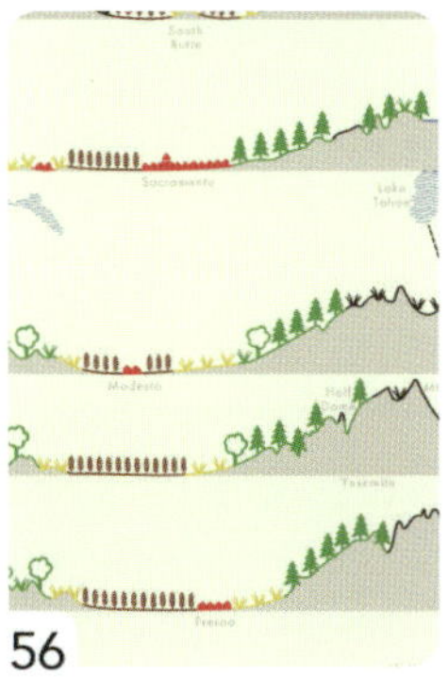
56

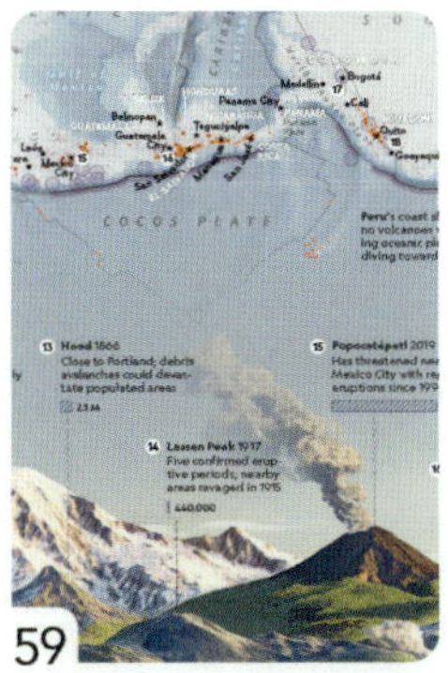

59

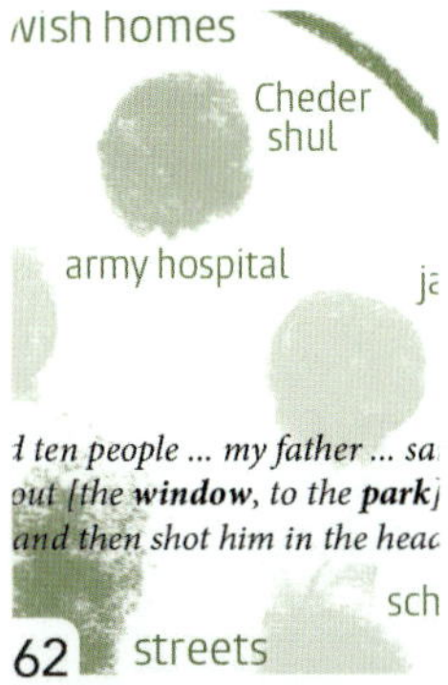

62

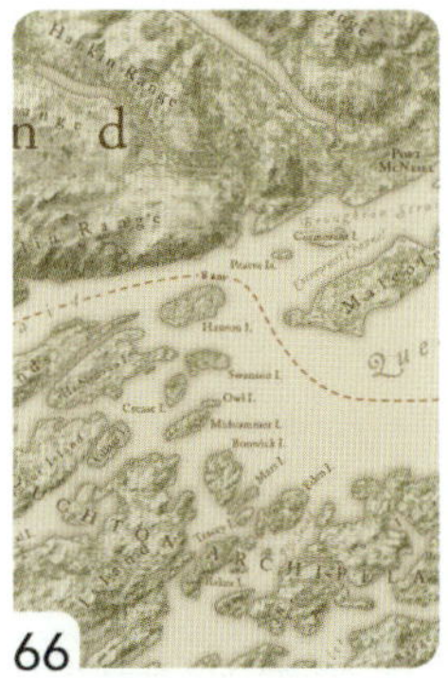
66

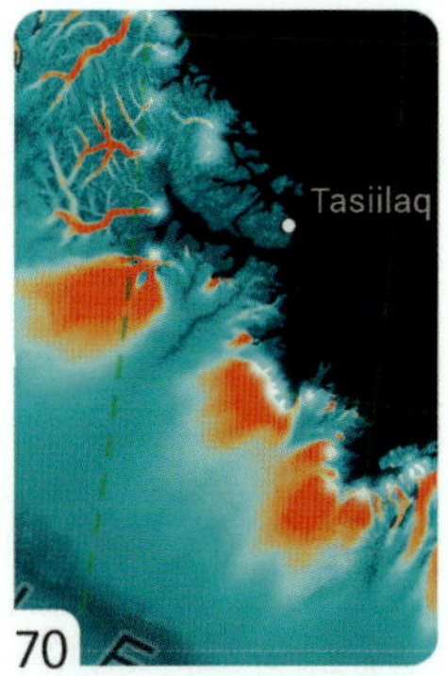

70

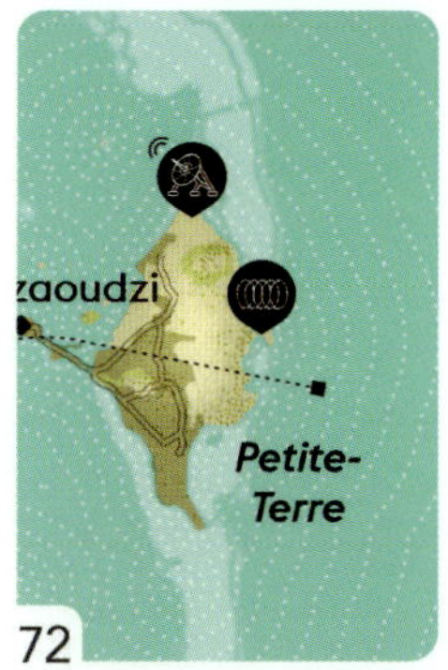

72

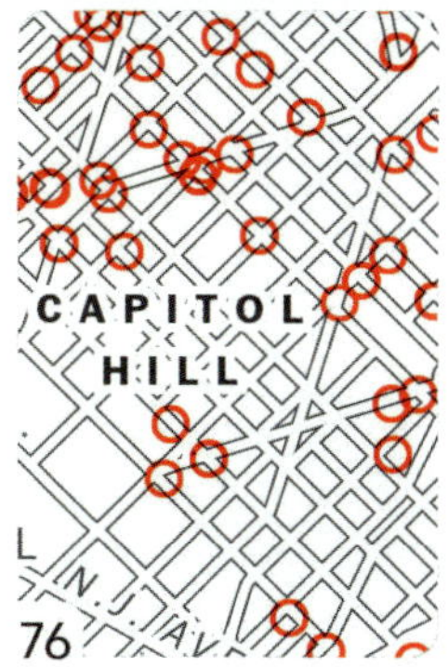

76

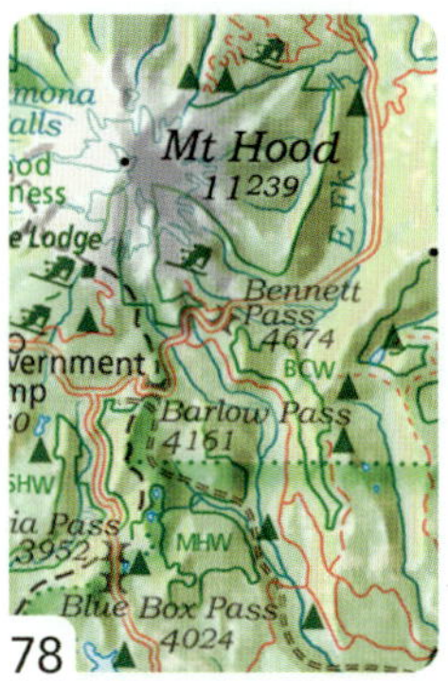

78

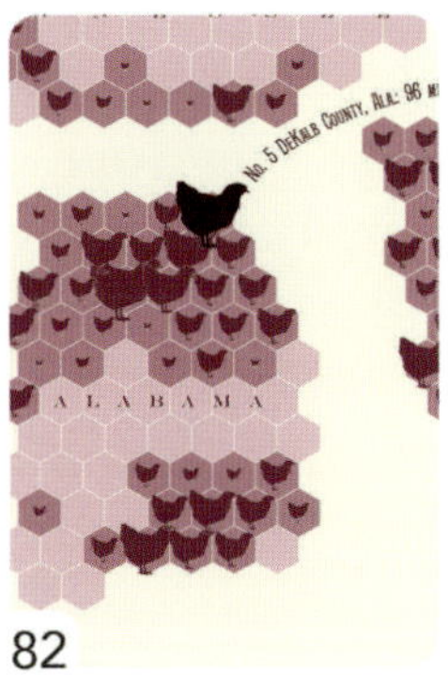

82

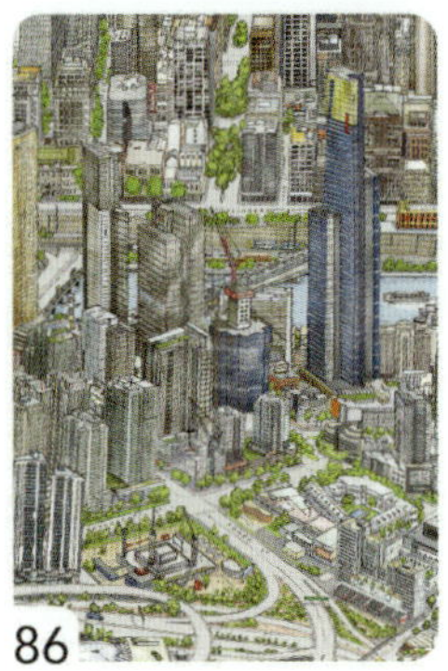
86

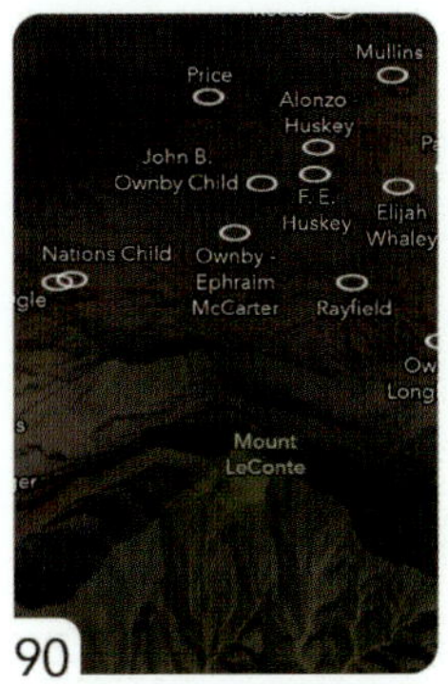

90

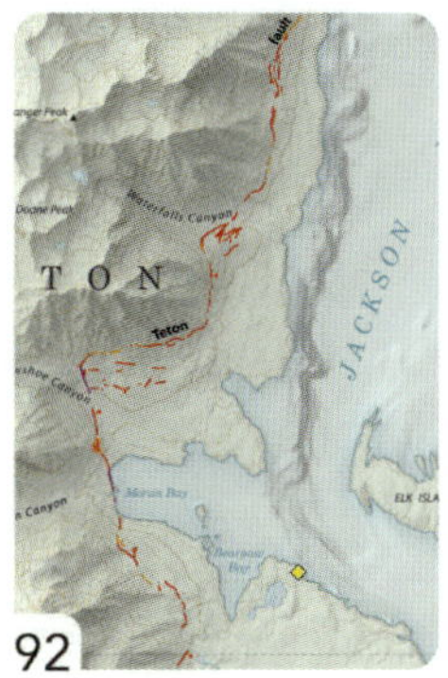

92

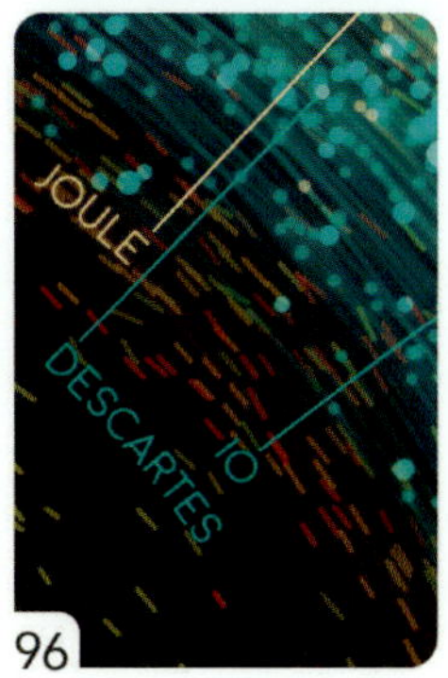

96

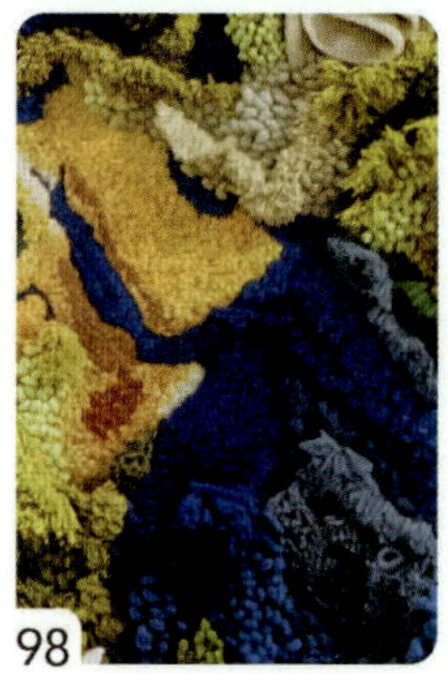
98

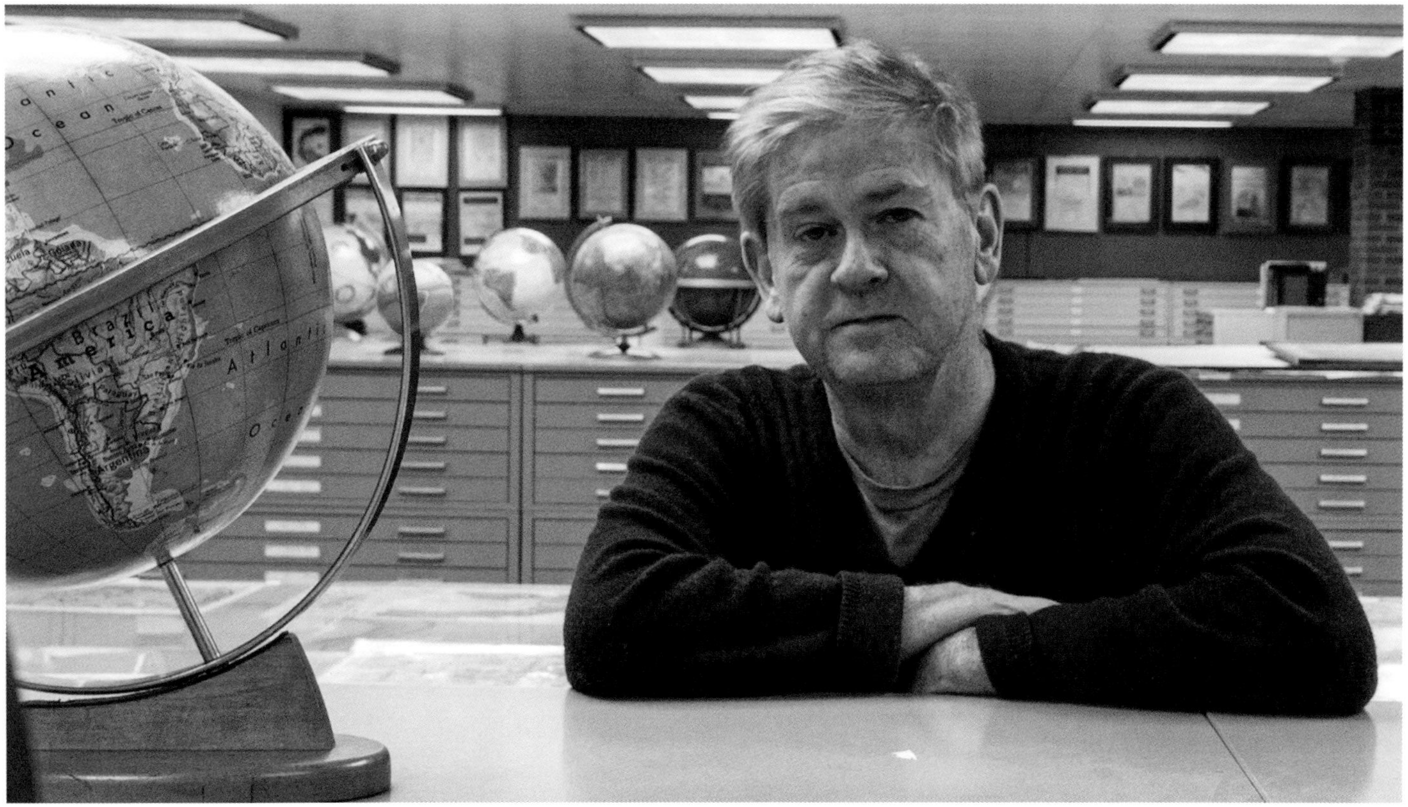

A Note from the President of NACIS

Dear Reader,

Thank you for your interest in this book you are holding, the fifth volume of *Atlas of Design*. As you look through the extraordinary maps presented in this book, you are contributing—wittingly or unwittingly—to one of the main goals of the North American Cartographic Information Society (NACIS): promoting communication among producers, distributors, and users of cartography.

Now celebrating its fortieth anniversary, NACIS has a proud history of supporting initiatives to further that goal. Its annual meetings are anchored by a conference featuring presentations offered in tracks on a wide variety of topics, and include Practical Cartography Day—a preconference event where tips, tricks, and techniques on mapmaking are offered—and a Map Gallery where members show off their work. But first and foremost, these meetings are social events where cartographers and cartographic enthusiasts exchange ideas in a genial atmosphere.

NACIS invests in its future by sponsoring two student competitions in map design and an undergraduate student scholarship, and by providing grants to attend the society's meetings. It assists the community by supporting Natural Earth, a rich map dataset in the public domain, and by making conference presentations freely available on the NACIS YouTube channel. It also publishes *Cartographic Perspectives*, an open-source journal of articles demonstrating creative research in cartography and geographic visualization. NACIS honors imaginative cartography through the Corlis Benefideo Award.

Finally, NACIS sponsors the *Atlas of Design*, a biennial publication showcasing the sublime art of mapmaking. As you peruse the maps in the volume you hold, it's my hope that you'll be inspired by what cartography can achieve. I invite you to visit our website, *nacis.org*, to learn more about NACIS and find details on the initiatives mentioned above. And to join us as we develop new initiatives.

–Leo Dillon

A Note from the Editors

When we agreed to edit Volume 5 of the *Atlas of Design* in spring 2019, we knew we were living in, as the old curse has it, "interesting times," but we had no idea just how "interesting" 2020 would be.... As we write this in fall 2020, we still don't know how the rest of this year will turn out. The North American Cartographic Information Society (NACIS) celebrates its fortieth anniversary in 2020, and we had hoped to hand out copies of the book to buyers at the annual meeting in Minneapolis and share thoughts about it with attendees. Instead, because of COVID-19, NACIS is holding its first virtual meeting, and the copy of the book you hold in your hands has been delivered by mail, not in person. So, whenever and wherever you are reading this, welcome! We hope you enjoy these maps made during what has been dubbed, half-jokingly, the "before times."

Like the four preceding volumes, Volume 5 presents a vast and diverse range of maps: technical works that communicate precisely measured data, on the one hand, and flights of fancy, on the other; work rendered digitally from beginning to end, and work composed of tangible materials by the creator's hand; maps designed for large organizations, and maps made for the amusement of their creators; maps of the solar system or the world, and maps of a tiny piece of it. We are grateful to the cartographers and artists who have allowed us to share their work with the world.

This book reflects NACIS's commitment to being a collaborative, volunteer-based organization: the judges, editors, and contributors, like the board and officers of the organization, offer this volume as a labor of love.

A record 450 maps were submitted for consideration in Volume 5. We are grateful to all those who submitted their work, for their interest and enthusiasm. We narrowed the field to 150 maps, which we forwarded to our stalwart team of judges, to whom we wish to extend a heartfelt thank-you: Anton Thomas, Daniel Huffman, Ian Muehlenhaus, Jim Meacham, Juliane Cron, Lauren James, Lauren Tierney, Madison Draper, Meghan Kelly, Tim Wallace, Tom Patterson, and Vicky Johnson.

We are grateful to our predecessors on the editorial team—Daniel Huffman, Tim Wallace, Sam Matthews, Marty Elmer, Ginny Mason, Lauren Tierney, and Alethea Steingisser—who set the bar high, gave us an excellent book design to build on, and provided guidance and support throughout.

Our copyeditors for Volume 5, Lys Weiss and Jeffrey K. Weiss of Post Hoc Academic Publishing Services, helped keep our language consistent and on track, while preserving the distinctive voice of each contributor. Shapco Printing, a new provider for this volume, has steadfastly supported turning our vision for Volume 5 into a reality. East View Map Link has continued to make order fulfillment simple and straightforward.

Finally, we thank our readers, whose interest in cartography and design has driven this project for ten years and five volumes so far. We hope the mapmakers among you find inspiration for your own work and consider becoming part of the next volume. Thank you for supporting the *Atlas of Design*, and may all map enthusiasts find this as fascinating a collection of work as we do.

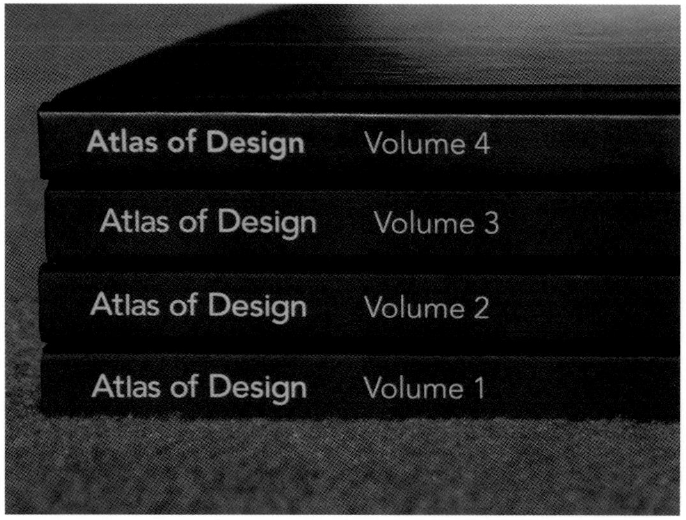

Introduction

As editors, we are struck by the variety of production and design technologies that went into the maps in this volume. Design of any kind is constrained by the technologies available to execute it. As with all craftspeople, our work bears the marks of our tools. But the variety of technologies used got us thinking and asking questions, such as: How do the tools we use and the frameworks we work within affect us?

We asked several long-standing NACIS members for their thoughts on technologies used in their work, or the work they have studied. Here is what they had to say:

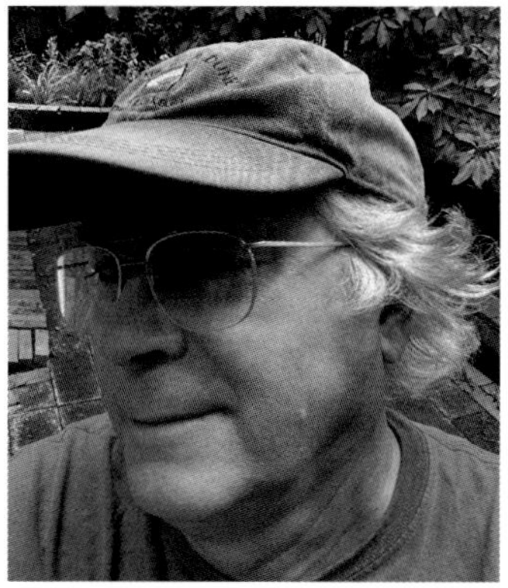

Tom Patterson

Having started my cartographic career in the 1970s, using pen and ink, it is no exaggeration for me to say that technology has had a seismic impact on my approach to design through the years. Back then, thinking carefully about a design and drawing pencil mockups was absolutely necessary before launching into map production, which was an entirely separate activity. The pace was much slower; delayed gratification was the order of the day. And if you messed up, design remakes were very costly.

Nowadays, design and production have become merged. The data drive everything. On a new map project, the first thing that I do is obtain all relevant data. I then throw these data on a base map, combine them every which way, and look for the spatial stories that they tell. If I am lucky, happy accidents happen. It is an unscripted process. I am in the throes of creativity, working at a feverish pace. I lose track of time. And it is why I am still addicted to making maps.

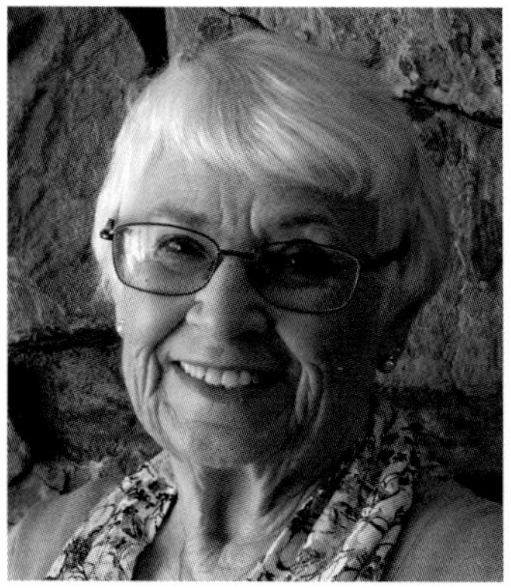

Judith Tyner

Technology has always affected cartographic work. Although we think now of technology as computers and GIS, this hasn't always been the case. The change from vellum to paper, the introduction of printing, the shift from intaglio to letterpress to lithograph, from hand lettering to mechanical lettering, from quill pens to ruling pens—all have had an impact on the design of maps, both in the sense of planning and the final "look" of the map. Sometimes it takes a while for the look to catch up to the technology. Early maps in the 1950s made with Leroy lettering look crude and mechanical now, just as early GIS maps do.

Many cartographic historians weren't interested in maps of the late nineteenthth century and certainly not those of the twentieth century because as maps became more scientific, they became "sterile" and less artistic.

I am not sure that there is a "standard" design process in cartography. And what do we mean by "design," anyway? Design can refer to the look of the map, the font and color, the symbols, even the projection. It can also refer to planning the map and making decisions of what font to use, whether color or black and white, and whether to use an equal area or conformal map projection. Each map or map series is a new design challenge.

When we ask how technology "affects the way people think about aesthetics or achieving results," are we talking about map users or mapmakers?

Generally I think most map users are not concerned with what technology is used to make the map, but rather with whether it is useful to them or, perhaps, is it attractive?

But I have met people who believed that GPS maps are beamed down from satellites in their entirety, and some who feel that maps made on a computer are more accurate than others "because human error isn't involved."

An issue for many cartographers is that many computer-generated maps seen in newspapers and journals violate basic rules of cartographic design. This isn't actually the fault of the technology, but of the maker. Many have not been trained in cartographic design and thus may create misleading maps and even "ugly" maps.

One issue I have had with the "new" technology is the assertion that it has democratized cartography. That now anyone can make a map. Throughout history people have made maps without being professional cartographers. Often these were sketches on a piece of paper, but they might also be highly professional-appearing maps used in articles and books. Mastering the technology does not mean mastering design.

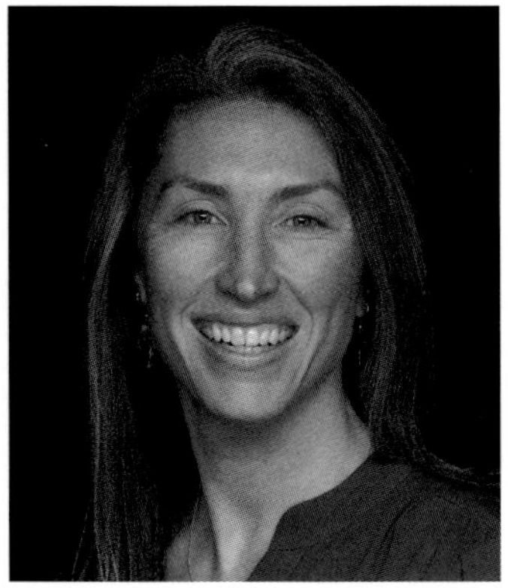

Rosemary Wardley

Digital cartographic tools are essential to my work and design process, especially in this time of remote work. From layout and conceptualization to research and production, I use almost all digital tools. Over the years, I have actually tried to step back from using technology for my initial design phase, as difficult as that is. I find that sketching out my ideas on paper and referring to physical books for design inspiration can be much more effective and efficient than starting on a computer right away. The first four volumes of the *Atlas of Design* have become invaluable to me, especially now without easy access to physical libraries and with only infrequent in-person interaction with colleagues.

As I'm sure is true in many fields, technology has allowed me to achieve much loftier goals and produce higher quality cartographic products than would otherwise have been possible. As someone who is not naturally gifted in drawing, I have often found that my aesthetic desires surpass my artistic abilities. The digital cartographic toolkit is helpful because it allows me to achieve more complex aesthetics despite not being a fine artist. And with so many fellow cartographers who readily share their processes and tools, it feels easier than ever to achieve the results you desire.

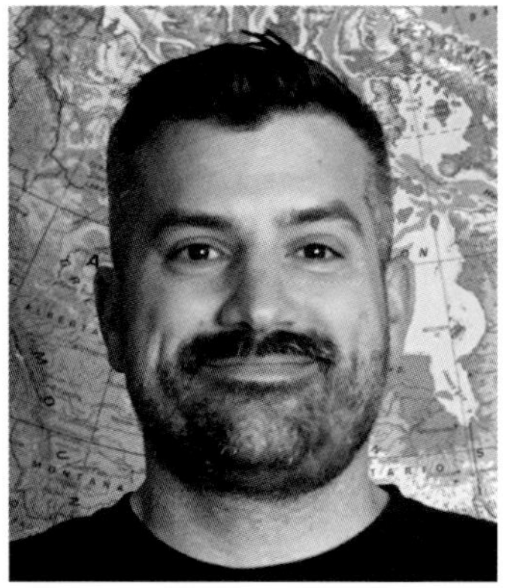

Jonah Adkins

Use of technology in my design process in work is usually driven by requirements. Clients and/or jobs routinely come with expectations of a deliverable in a specific form, a project file in a certain format, or a website making use of a specific technology. Requirements routinely determine time, cost, usability, and most important, my experience with the specified technology.

Access to free and open-source technologies has had the most impact on my process. Specifically, over the last ten years, technologies like cartographic programming languages, vector tiles, and web-based cartographic solutions, and the wide availability of geographic data have allowed me to rapidly develop and design maps that just weren't feasible earlier in my career. If a lot of these technologies were not open, I wouldn't have had the opportunity to learn, try things, break things, and learn again.

Having a wide base of knowledge across technology platforms informs my decisions with respect to aesthetics. Sometimes your ideas, trial and error, mockups, and so on can be limited by certain technologies. Some software just can't do what you want, or the barrier may just be "it's too hard"—continual self-education is an important key to my process.

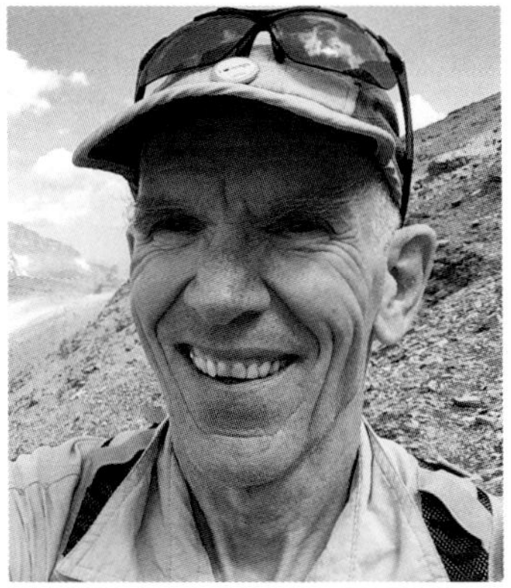

Steven R. Holloway

Inside the room. Technology dominates, defines, and controls the making of maps.

There is no time for being controlled by corporate lords, being servants of wealth and greed, participating in the destruction of the Earth. There is only time for one thing: saving the Earth. Be very, very certain your process of making has not imprisoned you within a seductive technology out of which you cannot see and hear.

Admit it, inside the room of mapmaking today—that is, digital post-postmodern cartography—it is hard to see outside the room. But there is an outside. And it is dying and under siege in part by the very technologies being employed on the inside.

Outside the room. Mapmaking comes to a standstill, there is only the place itSelf, and I am a part of that place. One Body. Listen, it may take some time. Does the place, the Body, request a map? Outside the room a map emerges as a Gift, not formed, defined, or servant to a technology but from the living body of the Earth. What emerges from this dialogue with a place is a Gift in the form of what might be called spatial relationships of interwoven and interacting information. There is *not* yet a technology or design associated or involved. There is, at this moment, no map, no design, no technology: just silence and the pressure that has emerged from within to respond. It may have taken several years to get to this moment, to hear the voice, to understand the necessity to make a response. The Gift is a pressure or voice to respond, to make visible.

Right MAP Making is a challenge. The voices I hear are gifts from the place itSelf, gifts not owned or shaped. The place itSelf informs what design, what technologies I employ. In my toolbox I have only a few technologies, and they are not in the current mapmaking room: the traditional printing technologies & matrices of stone lithography, relief, intaglio, collage, letterpress, book-making & silk-screen. Using hand-mixed inks, not always but often from gathered pigments with oils, my hands get dirty. There might be thread and water stains, handmade papers, chine-collé, torn and disassembled scraps of powders, and transparent bits of something else. The Gift determines the design. The Gift determines the technology. The Gift determines when and what to say. Should I find that my command of these technologies is insufficient to express the desire seeking expression, I probably won't/can't finish the making.

If [a map] feels like kicking over an ash can, a poet's castle, or a prosodic law, will you stop it? … If it happens to feel like trying to fly where humans cannot fly, to sing what cannot be sung, to walk in a cave on all fours, or to tighten up its girth in blind hope and faith and try to scale mountains that are not, who shall stop it? In short, must a [map] always be a [map] !?

— Charles Ives, Epilogue to *114 Songs*, 1961, replacing "song" with "map".

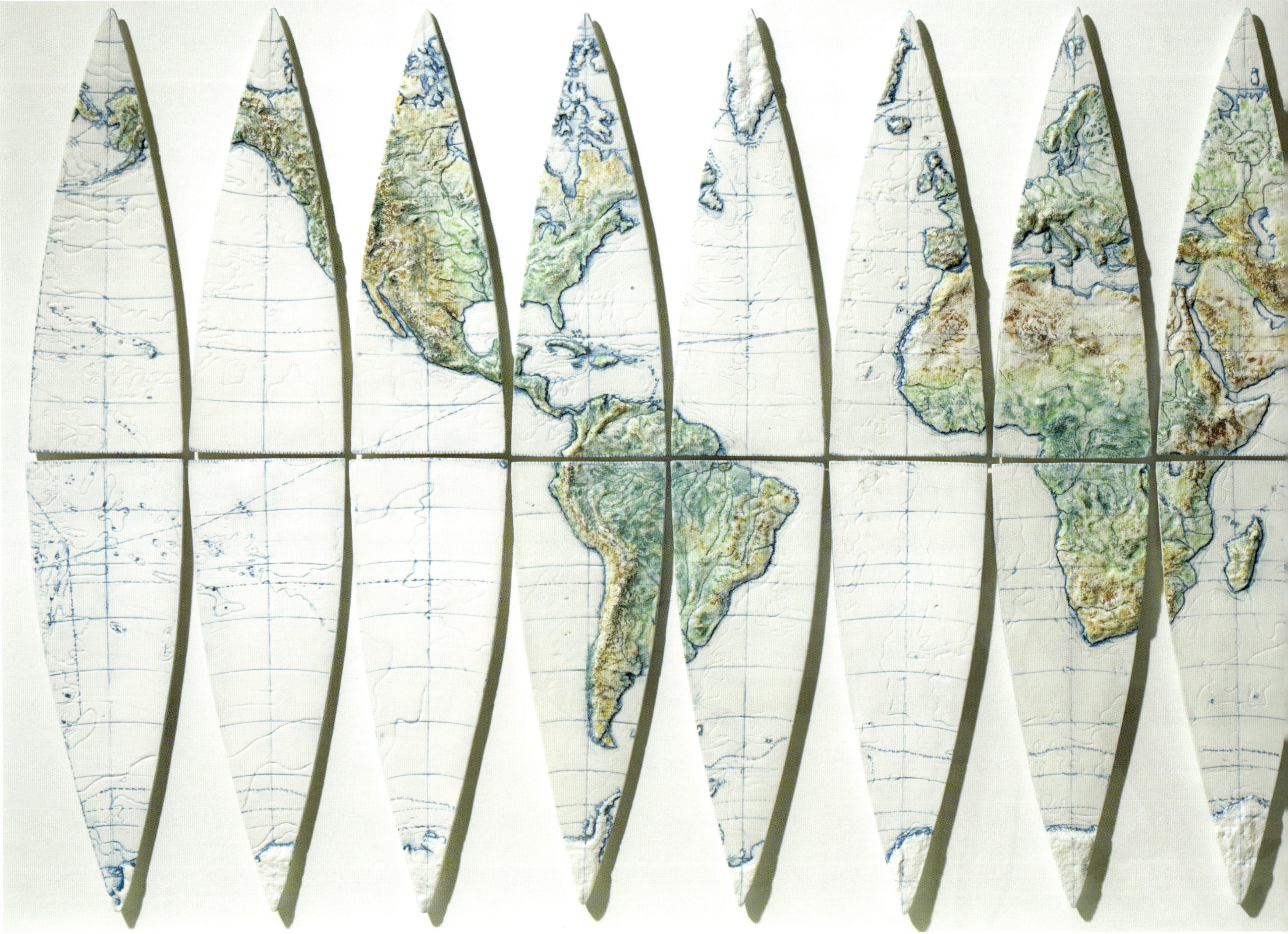

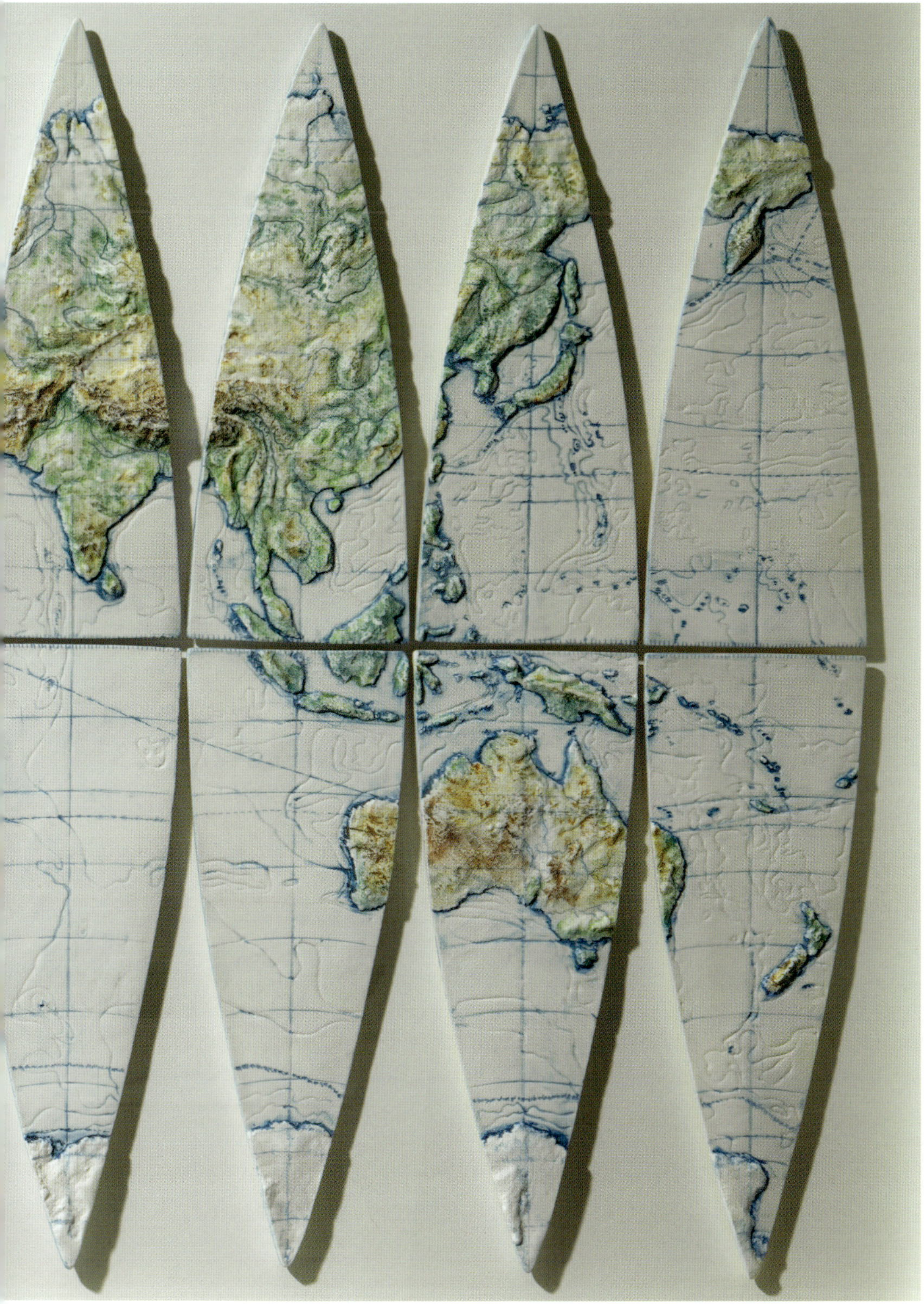

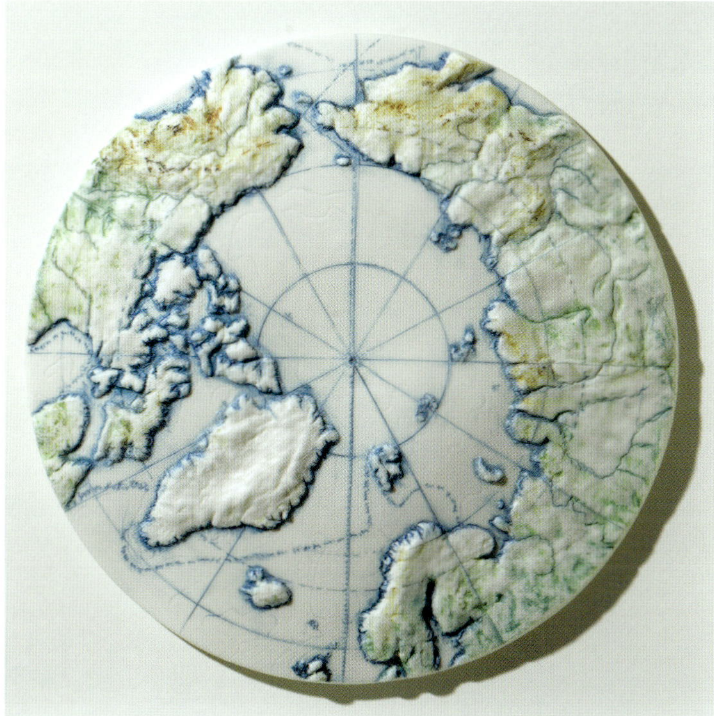

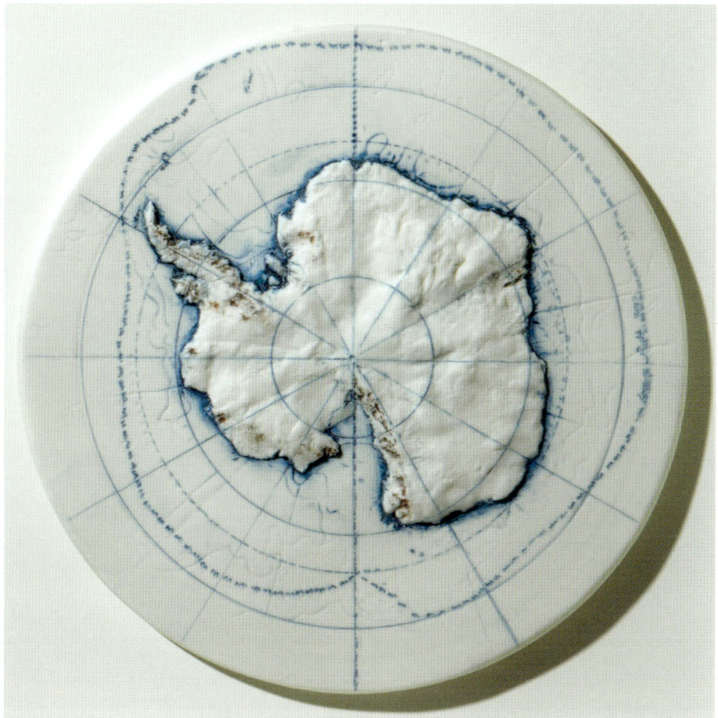

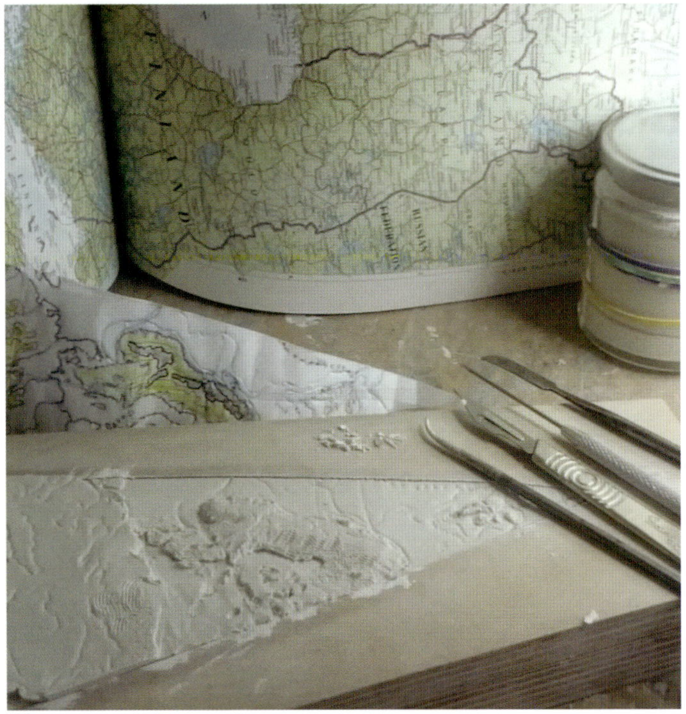

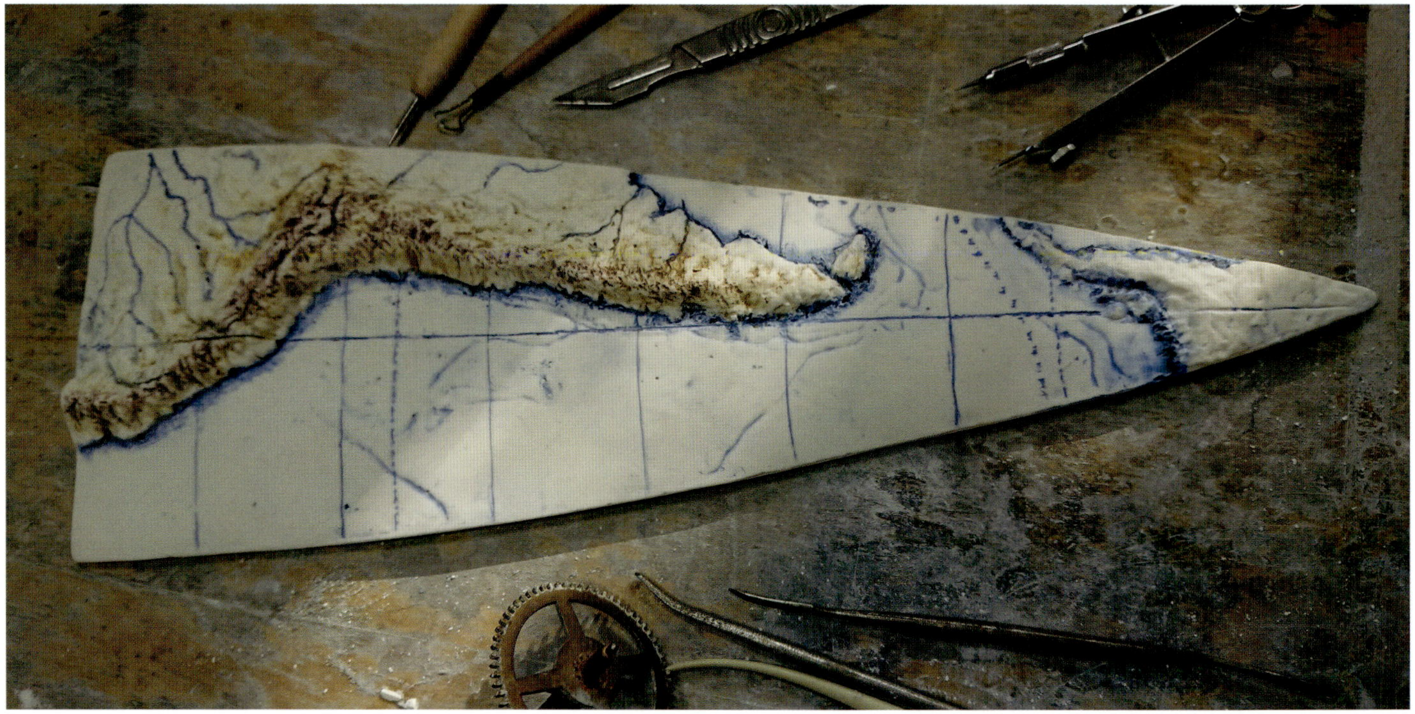

« Divided Globe Triptych: Segment and Poles Apart

Loraine Rutt

Given that my twin passions for cartography and clay formed in childhood, it was perhaps inevitable that I would combine the two in my work.

Divided Globe Triptych is a twenty-six-piece ceramic relief map. I first molded each tile in one-millimeter layers, forming a stepped relief model. From these, I sculpted relief by hand and eye. The malleable nature of porcelain lends itself well to depictions of natural topography; the fineness of porcelain particles can capture the tiniest of marks.

From plaster molds of the master tiles, I cast each edition in liquid clay "slip." I scribed the coastlines, graticule, and details by hand. After drying, I painted onto the raw clay, mindful that the metal oxide colors settle like river sediments, washing the porous surface like lapping tides. To finish, I fired the tiles at over 1200 degrees Celsius, fusing the clay and color particles.

Although I pursue attention to detail, I do not seek to create a replica. A theme that unites my work is human interaction with landscape. I combine a cartographer's exactitude with a desire to reveal more about the subject, using specific materials, processes, or treatments to document the impacts of human interaction with place over time.

Divided Globe Triptych is a 1,000-hour, knee-jerk response to the post-truth "fake news" permeating contemporary newsfeeds. It reflects the modern world where regions are separating, boundaries are being erected, and communities are increasingly divided.

This work counters the common perception of the world map as a rectangle. In a view shaped by Mercator's 500-year-old masterpiece, the Prime Meridian is centered, favoring Europe, while the North and South Poles barely register. Instead, I present the world in an equal-area projection, so the sizes of continents and oceans are accurate. Individually framed, the parts can be rearranged to place other locations at the center of the map.

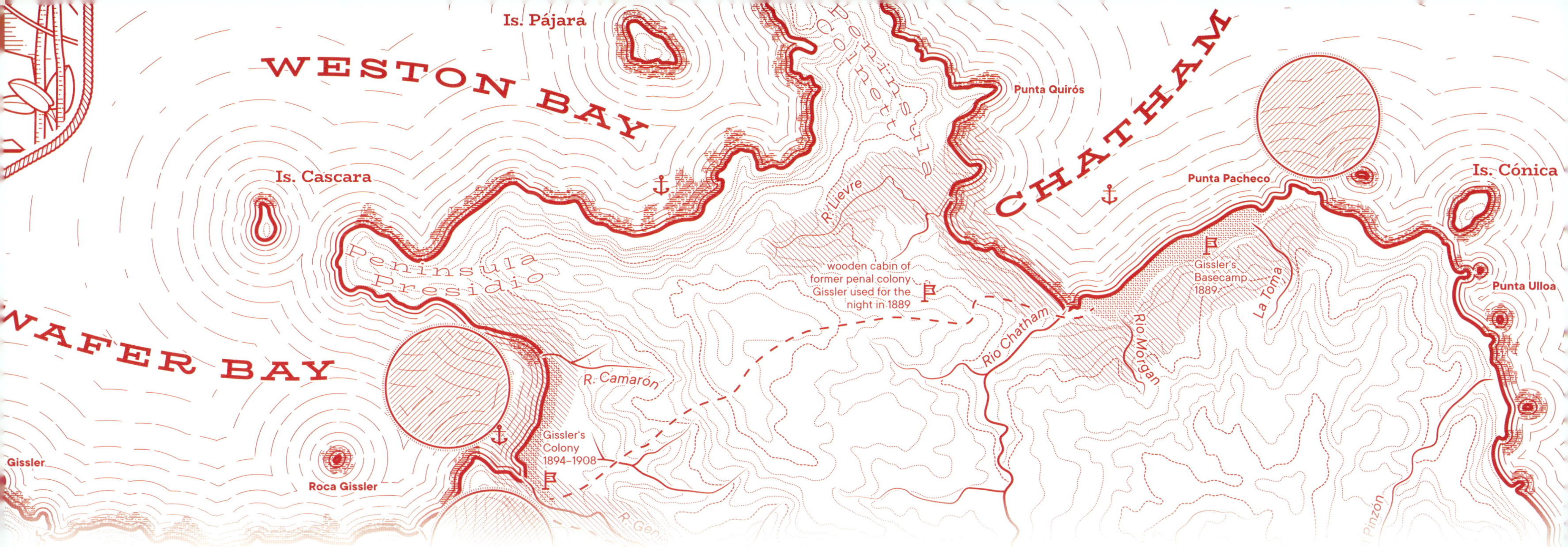

The Lost Treasures of Isla del Coco »

Jakob Listabarth

Maps can change one's life—and this is certainly true for August Gissler and the treasure maps of Cocos Island (Isla del Coco). Following rumors and real-life treasure maps, Gissler spent nearly twenty years (1889–1908) on this small, uninhabited island in the Pacific Ocean, looking for a mysterious treasure. The German-born Gissler (1857–1935) was so thoroughly convinced of the existence of treasures on Cocos Island that he founded a colony on the remote island with a few other German settlers. The Costa Rican government in 1897 named August Gissler the first and only "Gobernador de la isla del Coco" in history.

Treasure maps are fascinating, no doubt—even though their main actor, the desired treasure, is symbolized by nothing more than a mere letter X. That X represents the feeling of secrets, adventure, and anticipation. In this map I do not use an X; rather, I expand on the stories it embodies.

The monochrome representation, the graphic style, and even the specific color are inspired by a 1936 stamp featuring Cocos Island. The stamp was produced only one year after the death of August Gissler, making it a fitting visual reference for the time period.

The monochrome design approach forced me to think carefully about the use of patterns and symbols, and the visual hierarchy of map layers. The inset maps show where Cocos Island is situated and its role in the stories of hidden treasures. A series of small illustrations feature scenes of the island and key moments in its history.

This map continues the tradition of innumerable treasure maps of Cocos Island: for centuries, treasure hunters have attempted to recover the mysterious treasures suspected to have been hidden on the island—all with the help of treasure maps. Through the fine line work and a great degree of detail—despite the large format—I want to invite the reader to come a step closer, to really dive into the mysteries of the Isla del Coco.

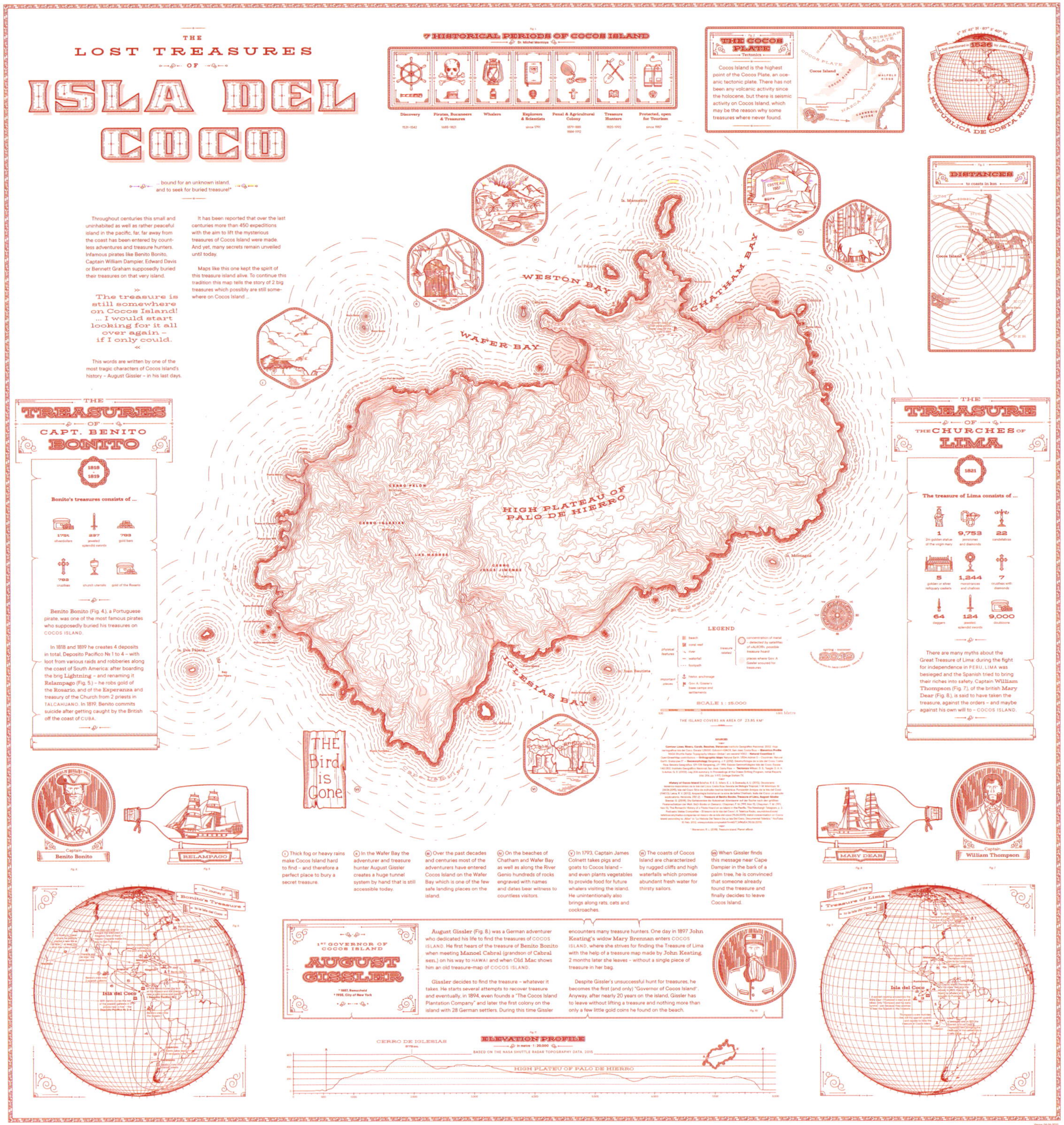

THE
LOST TREASURES
OF
ISLA DEL COCO
... bound for an unknown island, and to seek for buried treasure!*
Throughout centuries this small and uninhabited as well as rather peaceful island in the pacific, far, far away from the coast has been entered by countless adventures and treasure hunters. Infamous pirates like Benito Bonito, Captain William Dampier, Edward Davis or Bennett Graham supposedly buried their treasures on that very island.
It has been reported that over the last centuries more than 450 expeditions with the aim to lift the mysterious treasures of Cocos Island were made. And yet, many secrets remain unveiled until today.
Maps like this one kept the spirit of this treasure island alive. To continue this tradition this map tells the story of 2 big treasures which possibly are still somewhere on Cocos Island ...
» The treasure is still somewhere on Cocos Island! ... I would start looking for it all over again – if I only could. «
This words are written by one of the most tragic characters of Cocos Island's history – August Gissler – in his last days.
Fig. 1
7 HISTORICAL PERIODS OF COCOS ISLAND
Dr. Michel Montoya
Discovery 1531–1542
Pirates, Bucaneers & Treasures 1685–1821
Whalers
Explorers & Scientists since 1791
Penal & Agricultural Colony 1879–1881 1884–1912
Treasure Hunters 1825–1992
Protected, open for Tourism since 1987
Fig. 2
THE COCOS PLATE
Tectonics
Cocos Island is the highest point of the Cocos Plate, an oceanic tectonic plate. There has not been any volcanic activity since the holocene, but there is seismic activity on Cocos Island, which may be the reason why some treasures where never found.
COCOS PLATE
Cocos Island
COCOS RIDGE
CARIBBEAN PLATE
MALPELO RIDGE
NAZCA PLATE
CARNEGIE RIDGE
REPUBLICA DE COSTA RICA
DISTANCES
WESTON BAY
CHATHAM BAY
WAFER BAY
HIGH PLATEAU OF PALO DE HIERRO
IGLESIAS BAY
Is. Dos Pájaros
THE Bird is Gone
LEGEND
SCALE 1 : 25.000
THE
TREASURES
OF
CAPT. BENITO
BONITO
1818 - 1819
Bonito's treasures consists of ...
1751 silverdollars
237 jeweled splendid swords
783 gold bars
783 cruzifixes
church utensils
gold of the Rosario
Benito Bonito (Fig. 4.), a Portuguese pirate, was one of the most famous pirates who supposedly buried his treasures on COCOS ISLAND.
In 1818 and 1819 he creates 4 deposits in total, Deposito Pacifico № 1 to 4 – with loot from various raids and robberies along the coast of South America: after boarding the brig Lightning – and renaming it Relampago (Fig. 5.) – he robs gold of the Rosario, and of the Esperanza and treasury of the Church from 2 priests in TALCAHUANO. In 1819, Benito commits suicide after getting caught by the British off the coast of CUBA.
THE
TREASURE
OF
THE CHURCHES OF
LIMA
1821
The treasure of Lima consists of ...
1 2m golden statue of the virgin mary
9,753 jemstones and diamonds
22 candelabras
5 golden or silver reliquary caskets
1,244 monstrances and chalices
7 cruzifixes with diamonds
64 daggers
124 jeweled splendid swords
9,000 doubloons
There are many myths about the Great Treasure of Lima: during the fight for independence in PERU, LIMA was besieged and the Spanish tried to bring their riches into safety. Captain William Thompson (Fig. 7.), of the british Mary Dear (Fig. 8.), is said to have taken the treasure, against the orders – and maybe against his own will to – COCOS ISLAND.
Captain Benito Bonito
RELAMPAGO
MARY DEAR
Captain William Thompson
I Thick fog or heavy rains make Cocos Island hard to find – and therefore a perfect place to bury a secret treasure.
II In the Wafer Bay the adventurer and treasure hunter August Gissler creates a huge tunnel system by hand that is still accessible today.
III Over the past decades and centuries most of the adventurers have entered Cocos Island on the Wafer Bay which is one of the few safe landing places on the island.
IV On the beaches of Chatham and Wafer Bay as well as along the River Genio hundreds of rocks engraved with names and dates bear witness to countless visitors.
V In 1793, Captain James Colnett takes pigs and goats to Cocos Island – and even plants vegetables to provide food for future whalers visiting the island. He unintentionally also brings along rats, cats and cockroaches.
VI The coasts of Cocos Island are characterized by rugged cliffs and high waterfalls which promise abundant fresh water for thirsty sailors.
VII When Gissler finds this message near Cape Dampier in the bark of a palm tree, he is convinced that someone already found the treasure and finally decides to leave Cocos Island.
1ST GOVERNOR OF COCOS ISLAND
AUGUST GISSLER
* 1857, Remscheid
† 1935, City of New York
August Gissler (Fig. 8.) was a German adventurer who dedicated his life to find the treasures of COCOS ISLAND. He first hears of the treasure of Benito Bonito when meeting Manoel Cabral (grandson of Cabral sen.) on his way to HAWAI and when Old Mac shows him an old treasure-map of COCOS ISLAND.
Gissler decides to find the treasure – whatever it takes. He starts several attempts to recover treasure and eventually, in 1894, even founds a "The Cocos Island Plantation Company" and later the first colony on the island with 28 German settlers. During this time Gissler encounters many treasure hunters. One day in 1897 John Keating's widow Mary Brennan enters COCOS ISLAND, where she strives for finding the Treasure of Lima with the help of a treasure map made by John Keating. 2 months later she leaves – without a single piece of treasure in her bag.
Despite Gissler's unsuccessful hunt for treasures, he becomes the first (and only) "Governor of Cocos Island". Anyway, after nearly 20 years on the island, Gissler has to leave without lifting a treasure and nothing more than only a few little gold coins he found on the beach.
Fig. 10
Fig. 7
ELEVATION PROFILE
in metre · 1 : 20,000
BASED ON THE NASA SHUTTLE RADAR TOPOGRAPHY DATA, 2015
CERRO DE IGLESIAS
575 m
HIGH PLATEU OF PALO DE HIERRO
Bonito's Treasure
Isla del Coco
Treasure of Lima

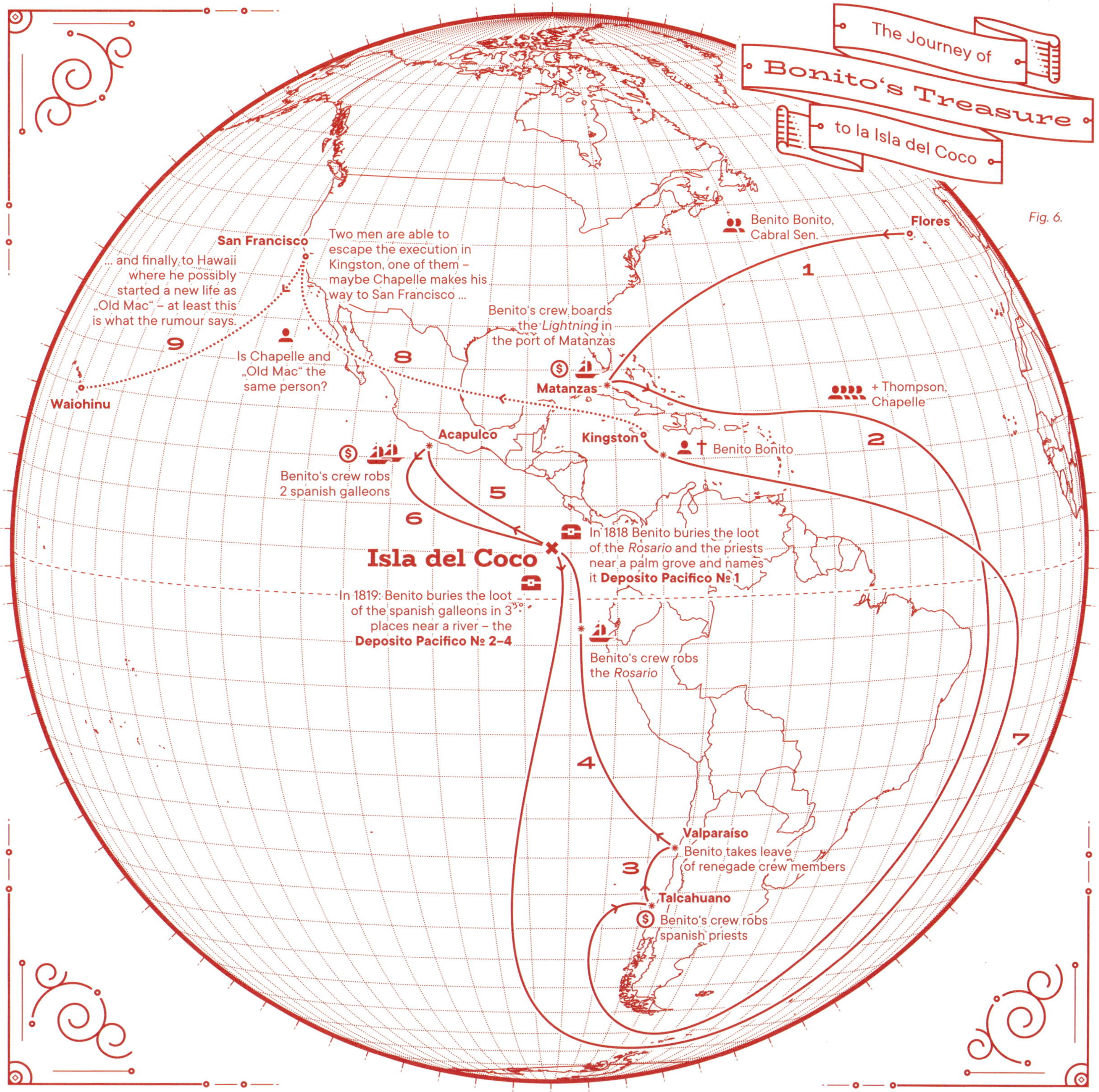
The Journey of
Bonito's Treasure
to la Isla del Coco
Fig. 6.
Benito Bonito, Cabral Sen.
Flores
1
+ Thompson, Chapelle
2
Benito's crew boards the Lightning in the port of Matanzas
Matanzas
Kingston
Benito Bonito
San Francisco
Two men are able to escape the execution in Kingston, one of them – maybe Chapelle makes his way to San Francisco ...
... and finally to Hawaii where he possibly started a new life as „Old Mac" – at least this is what the rumour says.
9
Is Chapelle and „Old Mac" the same person?
8
Waiohinu
Acapulco
Benito's crew robs 2 spanish galleons
5
6
Isla del Coco
In 1818 Benito buries the loot of the Rosario and the priests near a palm grove and names it Deposito Pacifico № 1
In 1819: Benito buries the loot of the spanish galleons in 3 places near a river – the Deposito Pacifico № 2–4
Benito's crew robs the Rosario
4
7
Valparaíso
Benito takes leave of renegade crew members
3
Talcahuano
Benito's crew robs spanish priests

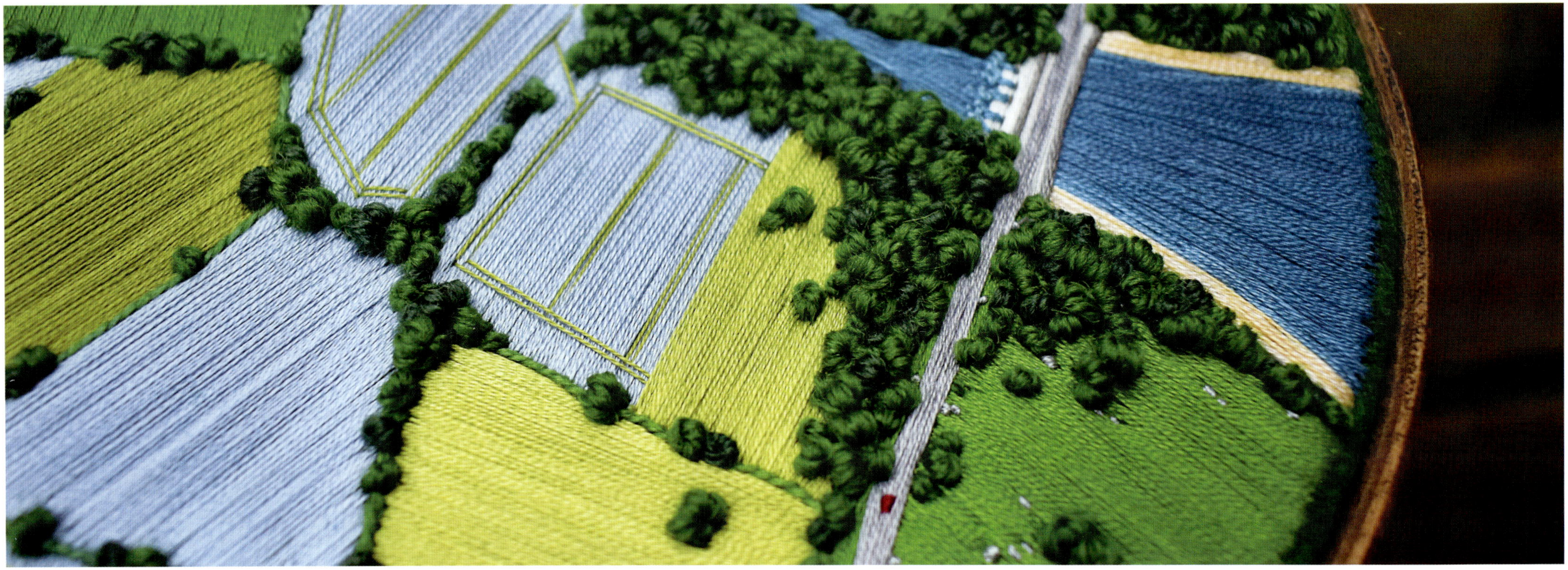

Embroidery »

Victoria Rose Richards

The Devonshire countryside of England is like a patchwork quilt. The way the natural and man-made aspects combine to create concise little patches in the land has always intrigued me: the agricultural fields that are both geometric and higgledy-piggledy, the snaking rivers and blooms of forest.... I wanted to re-create their forms and their textures in my favorite art medium, embroidery.

My aerial embroidery designs depict imagined landscapes based on the real English countryside. The county of Devon inspires most of these works, and some pieces are directly based on specific places near my home: for example, *Take a Drive on the Sunnyside* depicts a very long, straight stretch of road with patches of woodland that I drive down to get to the nearest town. Bright yellow flowering oilseed rape is grown there in the late spring, and I have fond memories of driving past those fluorescent meadows many times through my life!

I often start by adding trees for the forest, and let the shapes naturally come to me as I work. Using cotton thread, wool, and felt sheets in embroidery rings, I aim to create realistic textures and depth. Big, fluffy French knots create a lush, bobbly, almost cloud-like look that trees and forests have from above, while smooth, silky satin stitch mirrors the flat plains of farmland, and a rougher split stitch creates twisting country lanes.

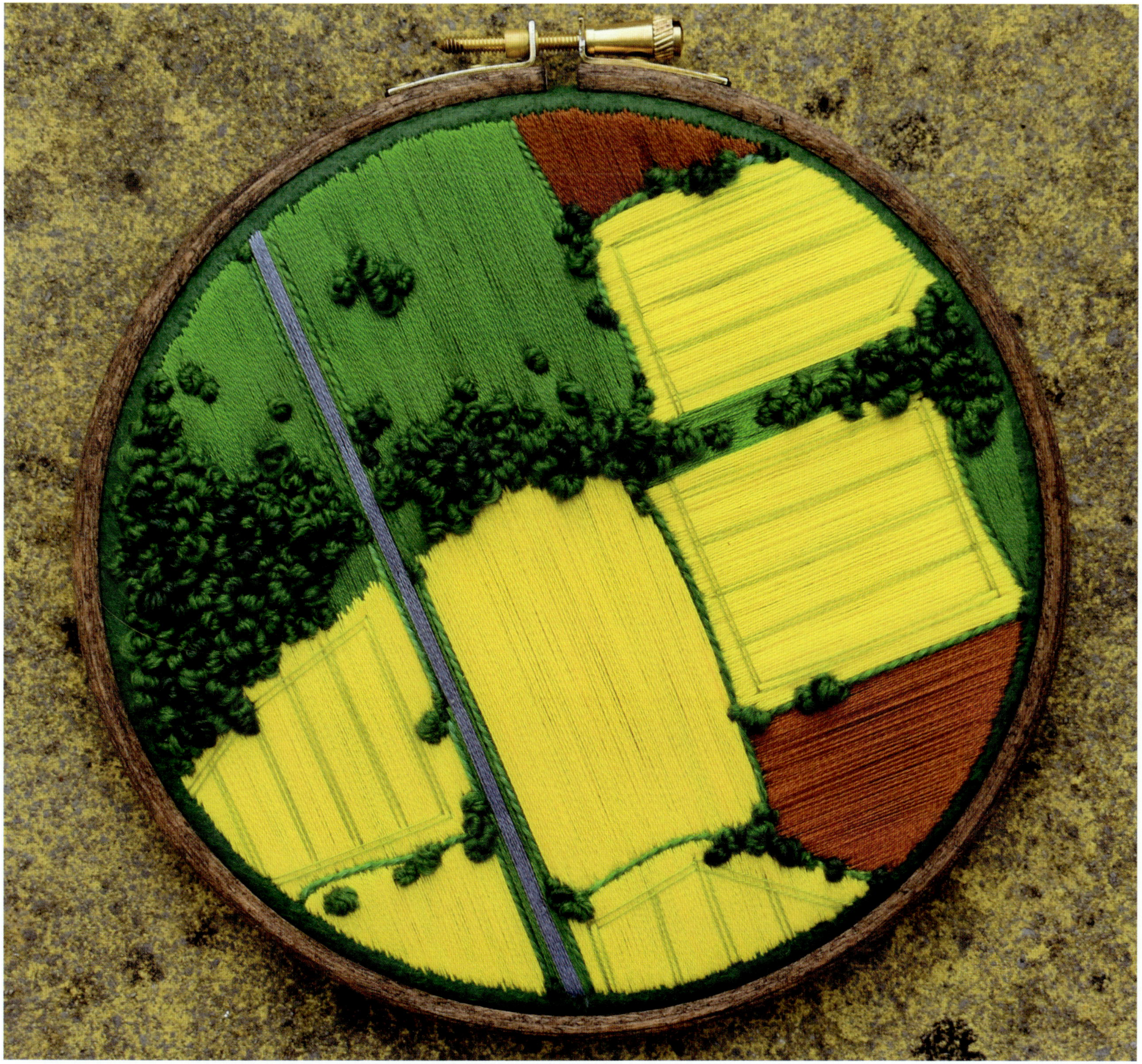
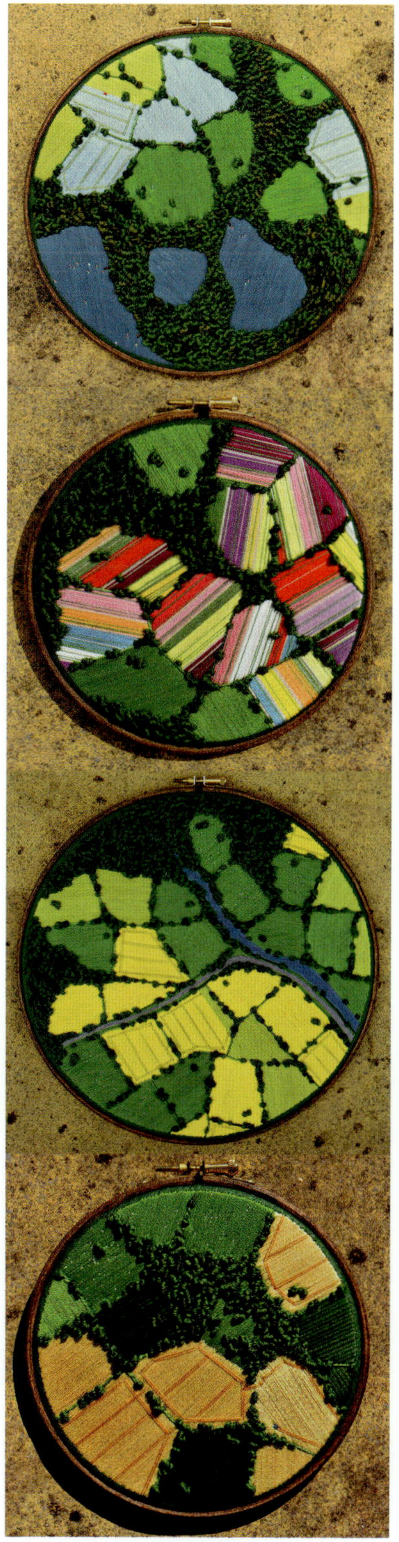

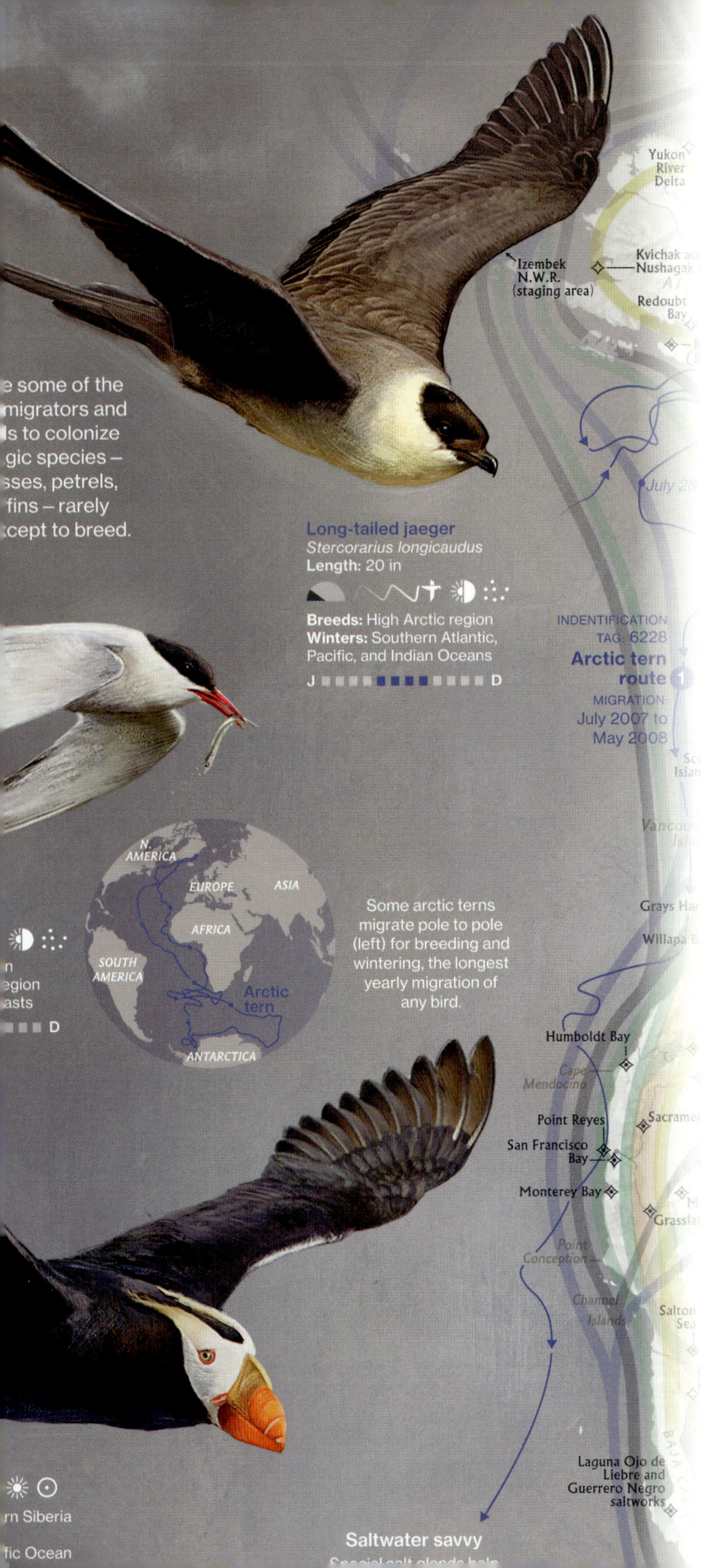

How Birds Migrate »

Fernando G. Baptista, Lauren E. James, Daisy Chung, Eve Conant, Irene Berman-Vaporis, Scott Zillmer, Adrienne Tong, Matthew W. Chwastyk, Clare Trainor, and Riley Champine — National Geographic

National Geographic magazine has a long history of mapping bird migratory patterns. Over the years several poster supplements have been distributed in the magazine. Continuing the tradition, this piece uses the latest tracking data to show migration patterns, with brand-new art created by Fernando Baptista integrated, illustrating notable species in scale. Using data from the Cornell Lab of Ornithology and Birdlife International, we mapped the migration schemes of five broad categories of birds, classifying them as seabirds, shore and wading species, birds of prey, waterfowl, and land birds.

Major flyways these long-distance travelers use to traverse the globe are shown as transparent lines, which fade and broaden to represent the way many species travel together and then disperse to an area. By using color that concentrates along the flyways and fades at the ends of the corridor, we symbolized the birds' behavior.

Symbols were developed to encode visually the time of migration of each species, their conservation status, and flight technique with formation. Notable species are richly illustrated by staff artist Fernando Baptista, all in their relative scale. Global coverage is achieved by featuring the Americas on one side of the poster, and the Eastern Hemisphere on the other.

Secondary illustrations feature wing design and the structure of feathers. Special care was taken to integrate the mapping and artwork to create a visually stunning and harmonious design. In particular, the map colors and visual feel take many cues from the appearance of the birds. Many hours were spent honing the design, shifting the placement of design elements and artwork to produce a harmonious, balanced, and engaging poster. Ultimately, hundreds of hours of research, design, and production went into the final product.

CELEBRATING THE YEAR OF THE BIRD

How Birds Migrate

Billions of birds migrate annually, crossing oceans, deserts, mountains, and even hemispheres in search of better weather and food. As this map of Western Hemisphere routes shows, these trips—which roughly half of all species make—can last up to several months, including time to rest and refuel. Physical adaptations often make these challenging journeys possible.

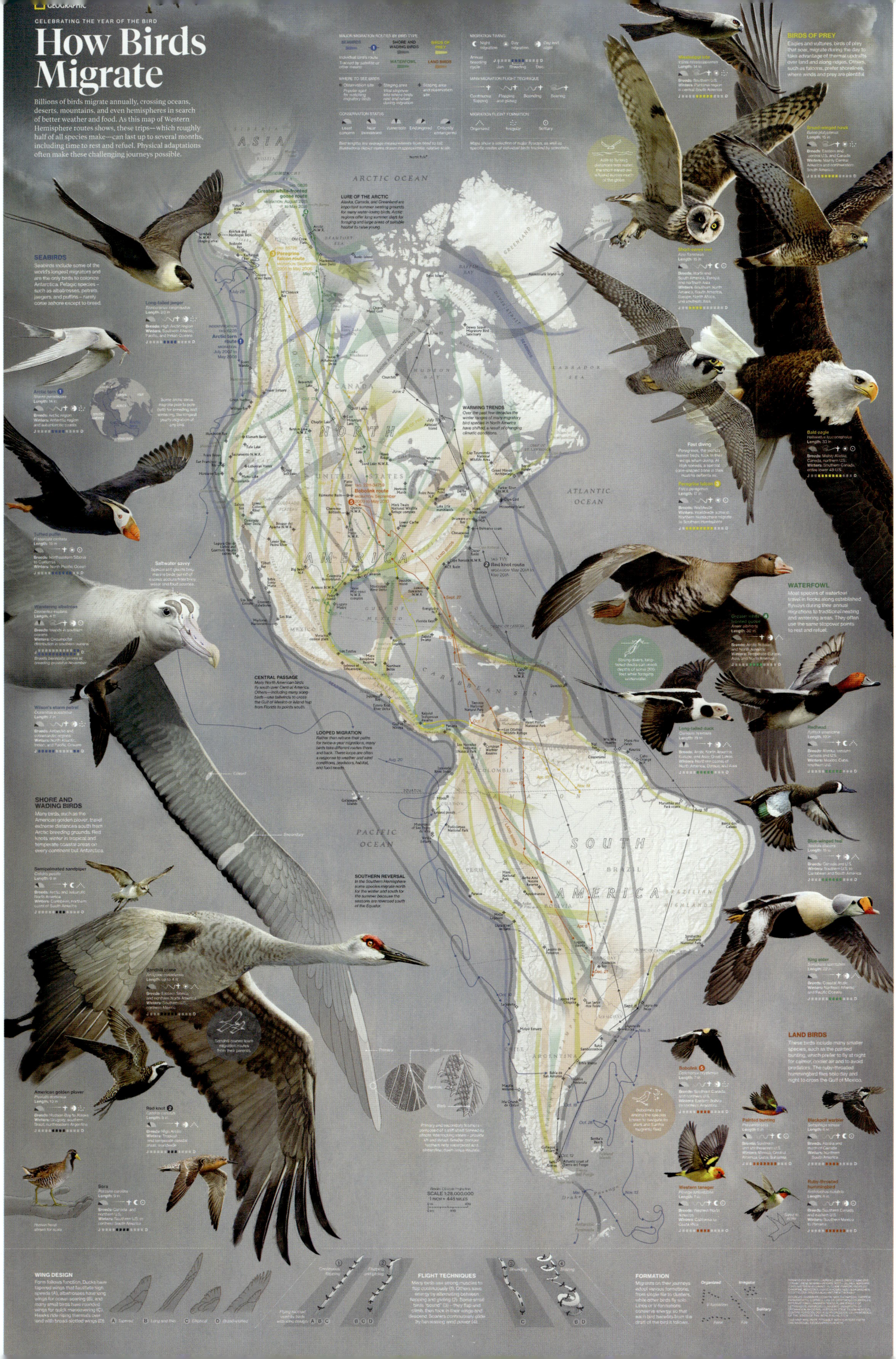

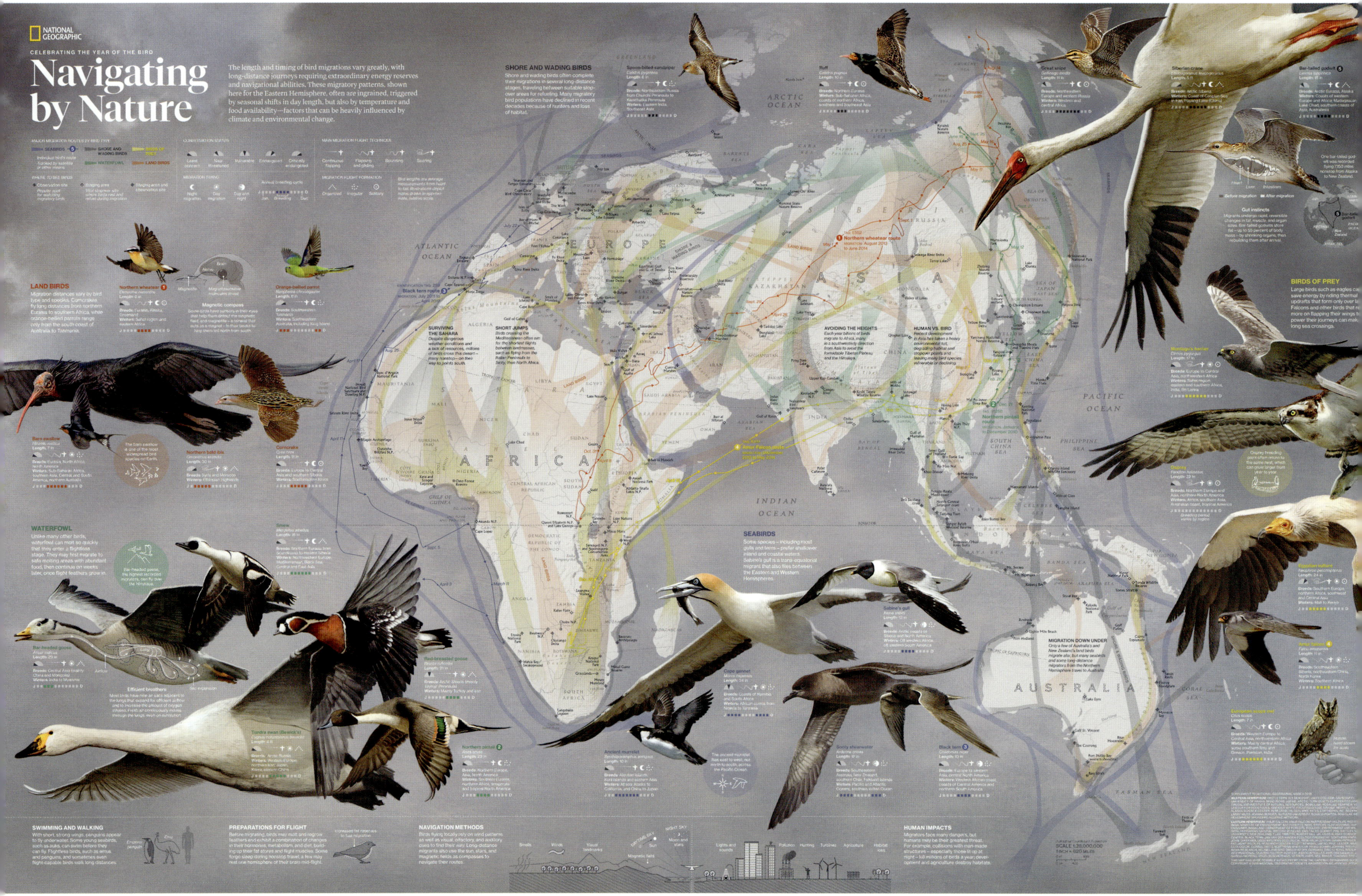
NATIONAL GEOGRAPHIC
CELEBRATING THE YEAR OF THE BIRD
Navigating by Nature
The length and timing of bird migrations vary greatly, with long-distance journeys requiring extraordinary energy reserves and navigational abilities. These migratory patterns, shown here for the Eastern Hemisphere, often are ingrained, triggered by seasonal shifts in day length, but also by temperature and food availability—factors that can be heavily influenced by climate and environmental change.
SEABIRDS
SHORE AND WADING BIRDS
BIRDS OF PREY
WATERFOWL
LAND BIRDS
Least concern
Near threatened
Vulnerable
Endangered
Critically endangered
Continuous flapping
Flapping and gliding
Bounding
Soaring
SHORE AND WADING BIRDS
LAND BIRDS
WATERFOWL
SEABIRDS
BIRDS OF PREY
ARCTIC OCEAN
ATLANTIC OCEAN
EUROPE
ASIA
AFRICA
INDIAN OCEAN
PACIFIC OCEAN
AUSTRALIA
SWIMMING AND WALKING
PREPARATIONS FOR FLIGHT
NAVIGATION METHODS
HUMAN IMPACTS

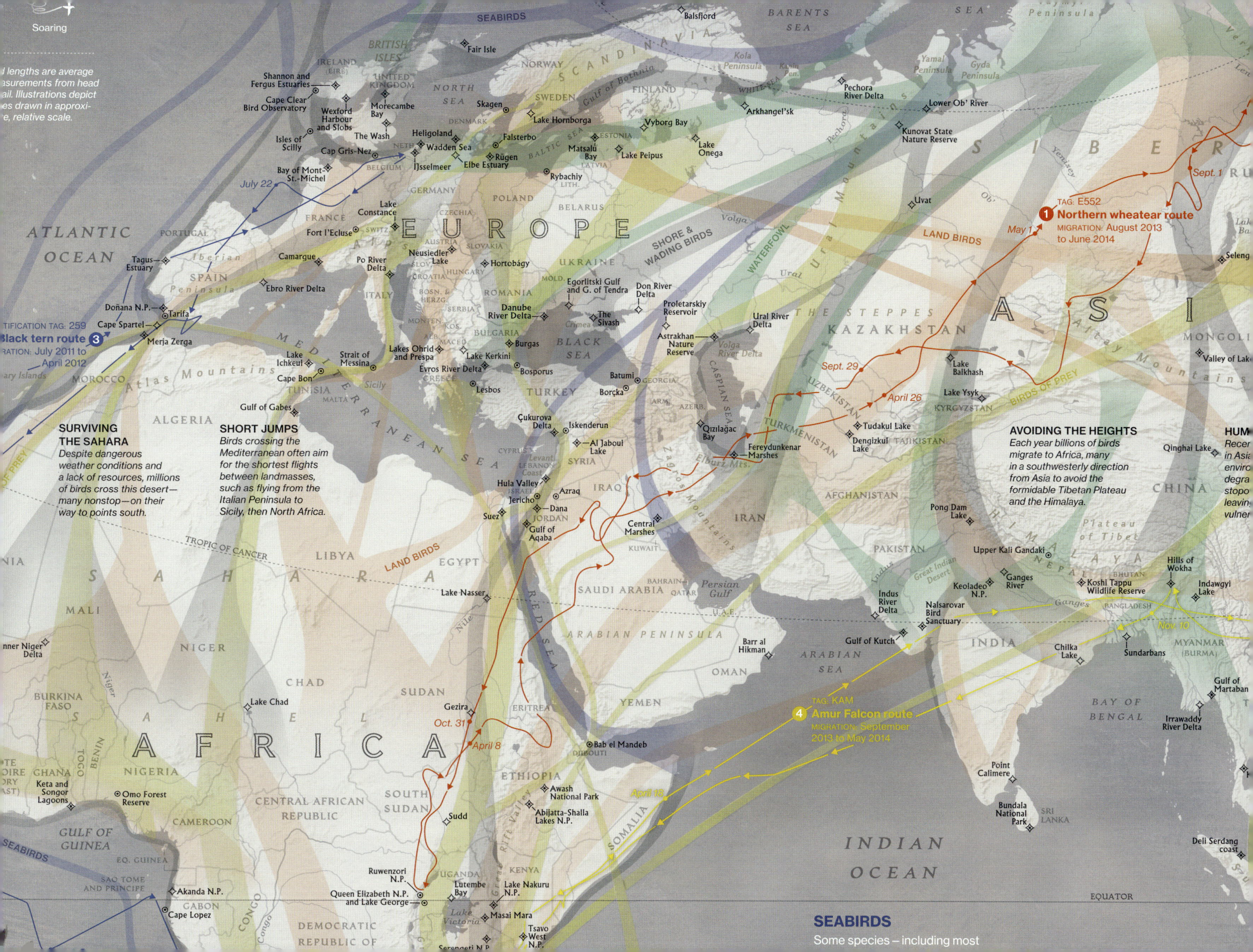
Soaring
d lengths are average
asurements from head
ail. Illustrations depict
es drawn in approxi-
e, relative scale.
SEABIRDS
BARENTS SEA
Balsfjord
Fair Isle
IRELAND (EIRE)
BRITISH ISLES
UNITED KINGDOM
NORWAY
SCANDINAVIA
Kola Peninsula
Yamal Peninsula
Gyda Peninsula
Shannon and Fergus Estuaries
Cape Clear Bird Observatory
Wexford Harbour and Slobs
Morecambe Bay
NORTH SEA
Skagen
SWEDEN
Gulf of Bothnia
FINLAND
WHITE SEA
Pechora River Delta
Lower Ob' River
Arkhangel'sk
Lake Hornborga
Vyborg Bay
Kunovat State Nature Reserve
Isles of Scilly
The Wash
DENMARK
Heligoland
Wadden Sea
Falsterbo
BALTIC SEA
ESTONIA
Matsalu Bay
Lake Peipus
Lake Onega
Cap Gris-Nez
Rügen
Elbe Estuary
IJsselmeer
LATVIA
Bay of Mont-St.-Michel
BELGIUM
Rybachiy
LITH.
July 22
GERMANY
POLAND
BELARUS
Uvat
SIBERIA
Yenisey
Sept. 1
TAG: E552
1 Northern wheatear route
MIGRATION: August 2013 to June 2014
May 1
ATLANTIC OCEAN
PORTUGAL
FRANCE
Lake Constance
CZECHIA
Fort l'Ecluse
SWITZ.
EUROPE
Volga
SHORE & WADING BIRDS
WATERFOWL
LAND BIRDS
Ural Mountains
Seleng
Tagus Estuary
Iberian Peninsula
SPAIN
Camargue
ALPS
Po River Delta
AUSTRIA
Neusiedler Lake
SLOVAKIA
Hortobágy
HUNGARY
UKRAINE
MOLD.
Egorlitski Gulf and G. of Tendra
Don River Delta
Ebro River Delta
CROATIA
ROMANIA
Proletarskiy Reservoir
Ural River Delta
THE STEPPES
KAZAKHSTAN
ASIA
Doñana N.P.
Tarifa
ITALY
BOSN. & HERZG.
SERBIA
Danube River Delta
Crimea
The Sivash
Astrakhan Nature Reserve
Volga River Delta
MONGOLIA
Valley of Lak
IDENTIFICATION TAG: 259
3 Black tern route
MIGRATION: July 2011 to April 2012
Cape Spartel
Merja Zerga
MEDITERRANEAN SEA
MONTEN.
KOS.
BULGARIA
Burgas
BLACK SEA
Lake Balkhash
Altay Mountains
Canary Islands
MOROCCO
Atlas Mountains
Lake Ichkeul
Strait of Messina
Lakes Ohrid and Prespa
MACED.
Lake Kerkini
Evros River Delta
GREECE
Bosporus
CASPIAN SEA
Batumi
GEORGIA
Sept. 29
UZBEKISTAN
BIRDS OF PREY
Cape Bon
TUNISIA
MALTA
Sicily
Lesbos
TURKEY
Borçka
ARM.
AZERB.
April 26
Lake Ysyk
KYRGYZSTAN
ALGERIA
Gulf of Gabes
Çukurova Delta
Iskenderun
CYPRUS
Qızılağac Bay
Tudakul Lake
Dengizkul Lake
TAJIKISTAN
TURKMENISTAN
Fereydunkenar Marshes
Elburz Mts.
Al Jaboul Lake
SYRIA
Levant
LEBANON
Coast
SURVIVING THE SAHARA
Despite dangerous weather conditions and a lack of resources, millions of birds cross this desert—many nonstop—on their way to points south.
SHORT JUMPS
Birds crossing the Mediterranean often aim for the shortest flights between landmasses, such as flying from the Italian Peninsula to Sicily, then North Africa.
AVOIDING THE HEIGHTS
Each year billions of birds migrate to Africa, many in a southwesterly direction from Asia to avoid the formidable Tibetan Plateau and the Himalaya.
HUM
Recer
in Asi
enviro
degra
stopo
leavin
vulner
Qinghai Lake
CHINA
Hula Valley
ISRAEL
Jericho
Azraq
IRAQ
Zagros Mountains
AFGHANISTAN
Pong Dam Lake
Plateau of Tibet
Suez
Dana
JORDAN
Gulf of Aqaba
Central Marshes
IRAN
KUWAIT
PAKISTAN
HIMALAYA
NEPAL
Upper Kali Gandaki
TROPIC OF CANCER
LIBYA
LAND BIRDS
EGYPT
SAHARA
BAHRAIN
QATAR
Persian Gulf
Indus
Great Indian Desert
Keoladeo N.P.
Ganges River
BHUTAN
Koshi Tappu Wildlife Reserve
Hills of Wokha
Indawgyi Lake
Lake Nasser
SAUDI ARABIA
U.A.E.
Indus River Delta
Nalsarovar Bird Sanctuary
Ganges
BANGLADESH
Nov. 10
MALI
Nile
RED SEA
ARABIAN PENINSULA
Gulf of Kutch
INDIA
Sundarbans
MYANMAR (BURMA)
Inner Niger Delta
NIGER
Barr al Hikman
OMAN
ARABIAN SEA
Chilka Lake
Niger
CHAD
SUDAN
BURKINA FASO
Lake Chad
ERITREA
YEMEN
TAG: KAM
4 Amur Falcon route
MIGRATION: September 2013 to May 2014
BAY OF BENGAL
Gulf of Martaban
Irrawaddy River Delta
Gezira
Oct. 31
AFRICA
April 8
Bab el Mandeb
DJIBOUTI
GHANA
TOGO
BENIN
NIGERIA
ETHIOPIA
Point Calimere
Keta and Songor Lagoons
Omo Forest Reserve
CENTRAL AFRICAN REPUBLIC
SOUTH SUDAN
Awash National Park
April 18
Abijatta-Shalla Lakes N.P.
Rift Valley
Bundala National Park
SRI LANKA
GULF OF GUINEA
CAMEROON
Sudd
SOMALIA
INDIAN OCEAN
EQ. GUINEA
SEABIRDS
SAO TOME AND PRINCIPE
Ruwenzori N.P.
UGANDA
KENYA
Deli Serdang coast
Akanda N.P.
GABON
Cape Lopez
Congo
CONGO
Queen Elizabeth N.P. and Lake George
Lutembe Bay
Lake Victoria
Lake Nakuru N.P.
Masai Mara
EQUATOR
DEMOCRATIC REPUBLIC OF
Serengeti N.P.
Tsavo West N.P.
SEABIRDS
Some species – including most

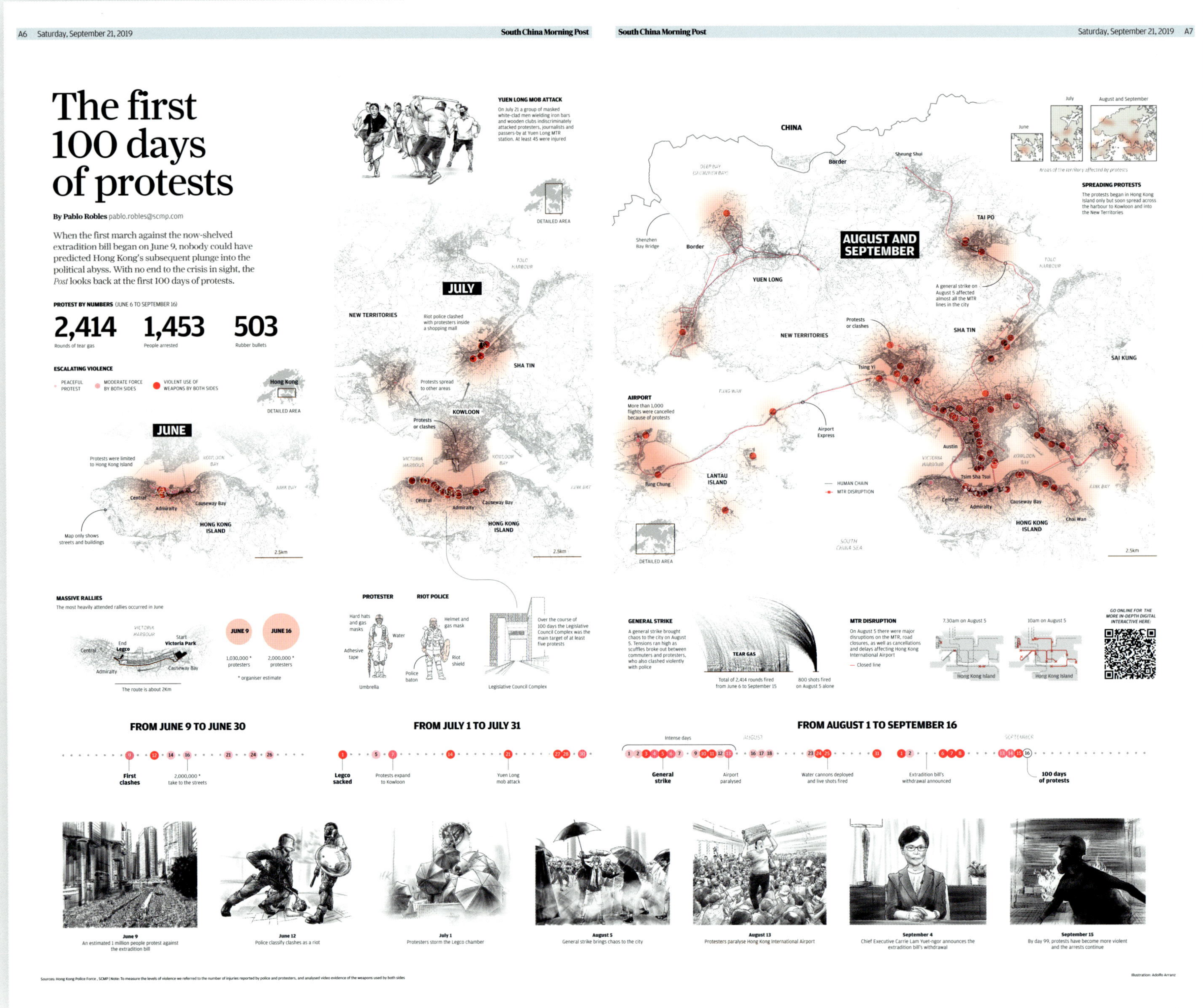

A6 Saturday, September 21, 2019 South China Morning Post | South China Morning Post Saturday, September 21, 2019 A7

The first 100 days of protests

By Pablo Robles pablo.robles@scmp.com

When the first march against the now-shelved extradition bill began on June 9, nobody could have predicted Hong Kong's subsequent plunge into the political abyss. With no end to the crisis in sight, the *Post* looks back at the first 100 days of protests.

PROTEST BY NUMBERS (JUNE 6 TO SEPTEMBER 16)

2,414 Rounds of tear gas

1,453 People arrested

503 Rubber bullets

ESCALATING VIOLENCE

June 9
An estimated 1 million people protest against the extradition bill

June 12
Police classify clashes as a riot

July 1
Protesters storm the Legco chamber

August 5
General strike brings chaos to the city

August 13
Protesters paralyse Hong Kong International Airport

September 4
Chief Executive Carrie Lam Yuet-ngor announces the extradition bill's withdrawal

September 15
By day 99, protests have become more violent and the arrests continue

Sources: Hong Kong Police Force , SCMP | Note: To measure the levels of violence we referred to the number of injuries reported by police and protesters, and analysed video evidence of the weapons used by both sides

Illustration: Adolfo Arranz

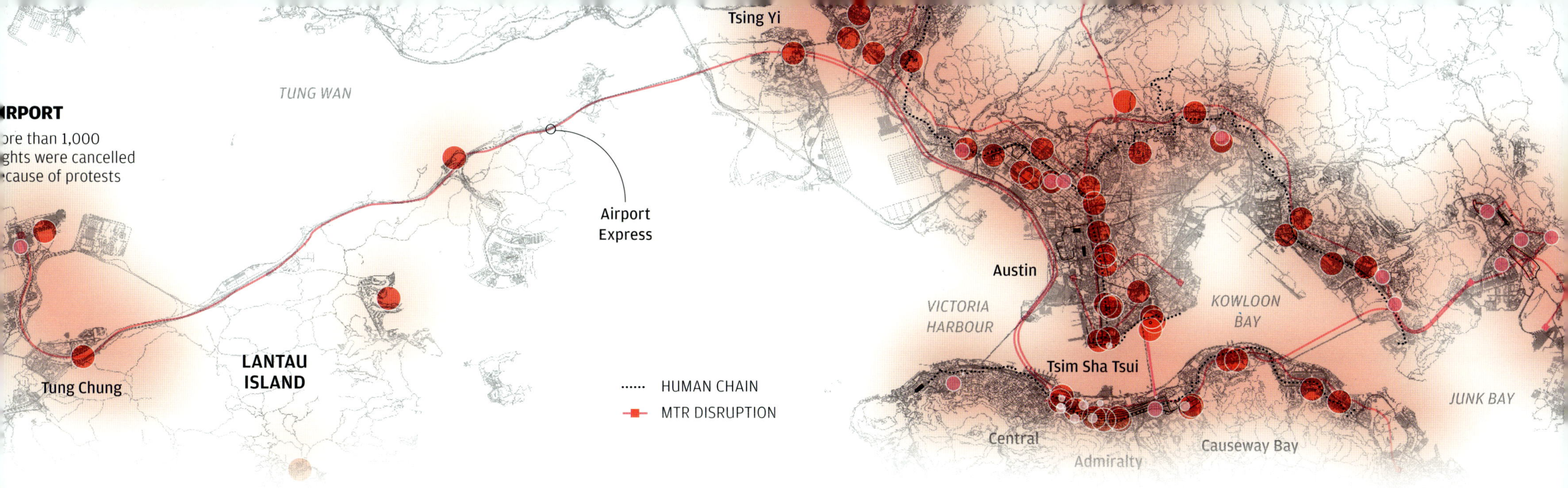

《 The First 100 Days of Protests in Hong Kong

Pablo Robles — South China Morning Post

The map was published in the *South China Morning Post* on Saturday 21, 2019. The newspaper has a print readership of around 340,000 people and an online presence of several million viewers.

Hong Kong's antigovernment movement sprang into life on June 9, 2019, when an estimated one million people marched against an extradition bill that was later withdrawn. As demonstrations gathered momentum, protesters adopted a "be water" philosophy of maximum disruption and adaptability, inspired by martial arts icon Bruce Lee. Protests were staged throughout the territory, sometimes occurring in multiple locations to catch police off guard.

The demonstrations became increasingly violent as protesters clashed with riot police and vented their frustration. There were acts of vandalism against the Mass Transit Railway, which some protesters accused of assisting police operations. Civilians with opposing political views also began to clash, most notoriously in the Yuen Long mob attack.

Three maps created in QGIS are placed across a double-page newspaper spread to illustrate how the protests evolved. By using shapefiles to indicate population density, streets, and affected train lines, the maps not only locate the major protests, but also indicate whether they were peaceful or violent.

The team manually collected geo-location data to compare police reports with dispatches from our frontline journalists to record how the protests spread across the entire territory and affected the daily lives of millions of Hongkongers. The maps include annotated notes about key events, such as general strikes, human chains, and the protest at Hong Kong International Airport, which resulted in more than three hundred canceled flights.

A black-and-white illustrated timeline with milestone events anchors the composition. To emphasize the data and the red highlights indicating where clashes occurred, a minimalist color palette was used. The red gradients also work as a visual metaphor by resembling welts and scars across the territory.

To learn more, a QR code links to our interactive story, which offers a deeper dive into the same events, which had the world transfixed. It combines more than twenty maps with dozens of photos and diagrams.

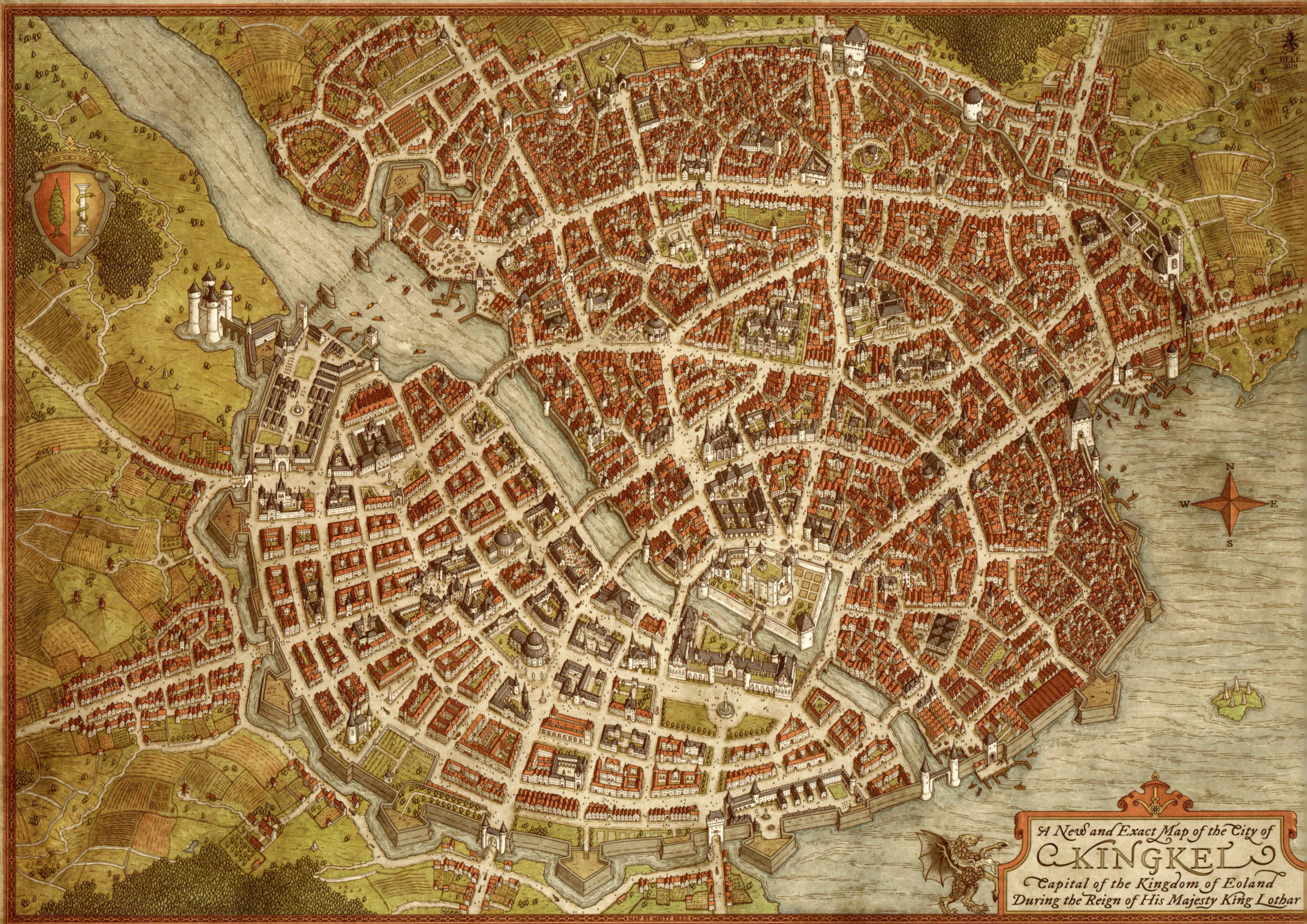
BEEE
2018
N
W
E
S
A New and Exact Map of the City of
KINGKEL
Capital of the Kingdom of Eoland
During the Reign of His Majesty King Lothar
MAP BY MISTY BEEE

« City of Kingkel

MistyBeee

Whether you notice or not, maps are all around us: in our phones, on websites, in travel books, at every bus stop. We count on them, and they're tools we all use at some point. When it comes to fantasy maps, this sense of reliability remains. Because we are accustomed to thinking that maps are trustworthy depictions of reality, facing the map of an imaginary world makes it tangible. I love to play with this ambiguity: the map, a serious, technical, useful object—transposed into imagination. A map is a gate.

Kingkel was one of my first city maps. Laura Poulain-Mair, the author for whom I drew it, needed an immersive visual to help her write her story, and she gave me complete freedom with regard to map style. My main inspirations are always the works of the greatest historical mapmakers, which, in this case, fit perfectly the requested spirit of the city. The fantasy style often reinvents history, and by my side I love to revisit the historic cartography, its immersive details, more artistic than scientific, and adapt it to our modern tastes and techniques, especially when it's about color and shading.

Regarding the structure of Kingkel, I wanted to distinguish two subsets: the northern and southern parts. The northern part, developed around the old castle, acquired progressively new districts while the city was growing, thanks to the trade coming from the nearby river and the lake. The newer southern part was planned, urbanized by a well-established government, offering its wealthy inhabitants all the comforts of "modern" architecture, and the security of improved walls.

Generally speaking, I think you can't build an imaginary map if you don't know the rules of the real world: how are mountains created, how do rivers flow, and so on. It's the same for cities: a city map will always be more consistent if you start with the reasons why people decided to live there. The first fort, the village, the little town… all leave their mark on the city in its golden age.

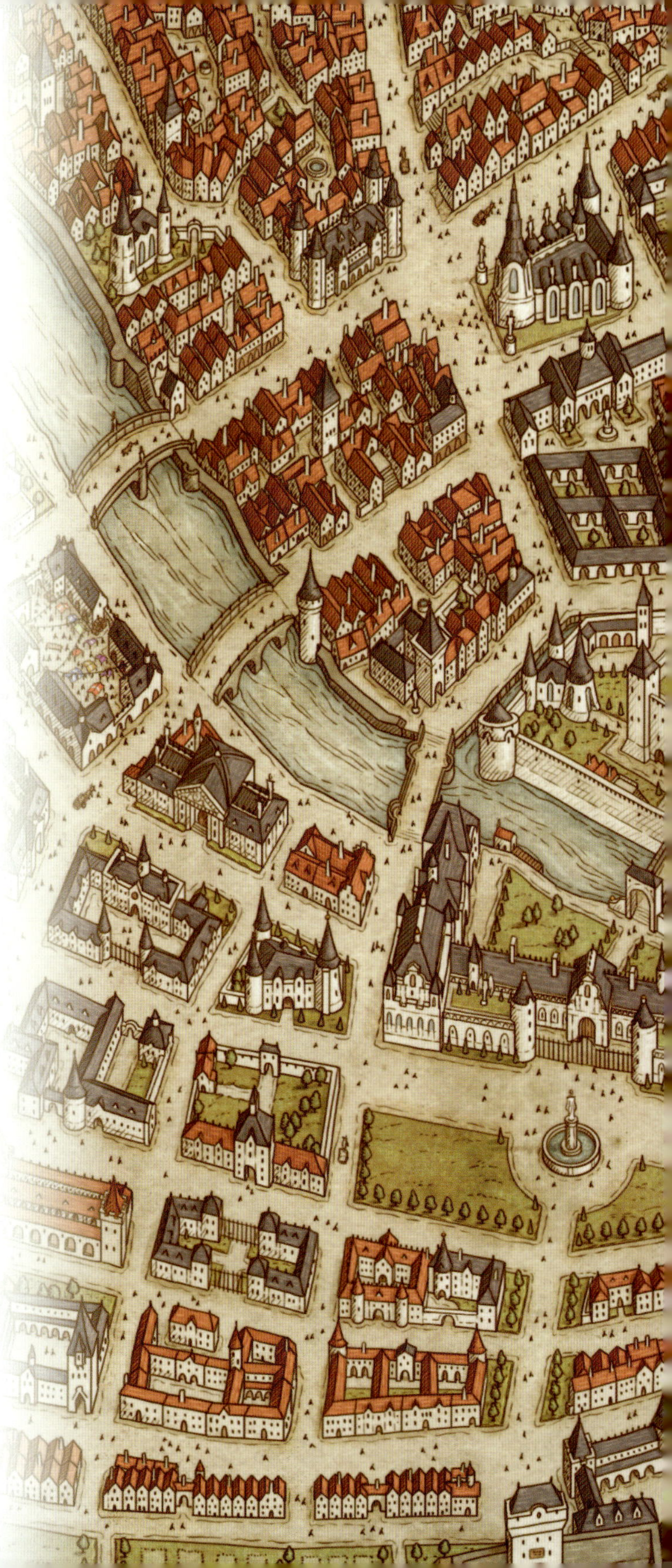

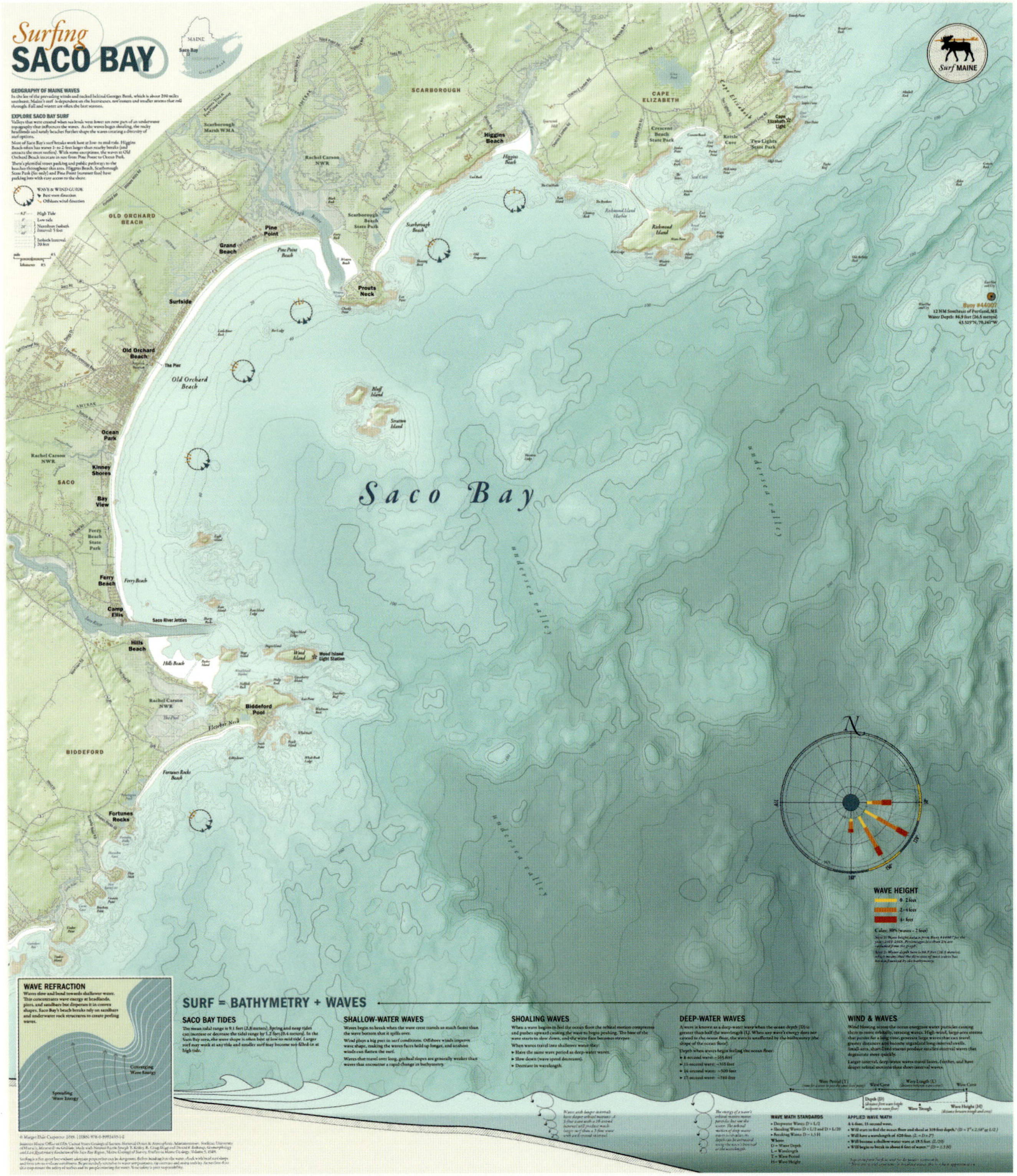

« Surfing Saco Bay

Margot Carpenter

Surfing Saco Bay began with my love of surfing and coastal topography. Where and when waves break is part of a much larger geography of wind, shoreline structure, and offshore bathymetry. I wanted to design a map that connected both surfers and coastal enthusiasts to this—not just a "here's a surf break" map, but one that begins to describe the complex coastal dynamics and invites the viewer to look beneath the waves.

The 11-mile arc of sandy coastline at Saco Bay, Maine is broken by rocky headlands and is adjacent to offshore valleys that were carved by rivers when sea levels were lower. The valleys help direct and form approaching waves. The locations and aspects of the rocky headlands, beaches, and river outlets influence what shape the wave will take when it arrives onshore.

It took many iterations to finalize a design palette. Differentiating the coastal topography was the most challenging part, but it allowed me to experiment with the type of colored line work often seen in early mountain cartography. Starting with aerial imagery and contour lines, I used denser, darker isolines to depict the rocky headlands; lighter, textured lines to show the beaches; and broken lines for near-shore bathymetry. For the water I chose the lighter, greener blues sampled from photos of backlit Saco Bay surf. This color range did a much better job of describing Saco Bay's shallower waters and allowed me to cleanly float text and graphics over it.

Many of the offshore valleys trend toward the northwest, the same direction as the traditional 315° lighting used when creating terrain relief shading, or hill shades. As a result, many of the valleys seemed to disappear. To offset this, I used multiple hill shades with varying lighting directions to shade and delineate the bathymetry. Depth contours helped organize the bathymetric shading and gave real numbers to the offshore terrain.

Graphic details, such as a wind rose with wave data replacing wind directions, and a bathymetric profile to illustrate wave dynamics further explain the connection between the bathymetry and the surf seen at the shore.

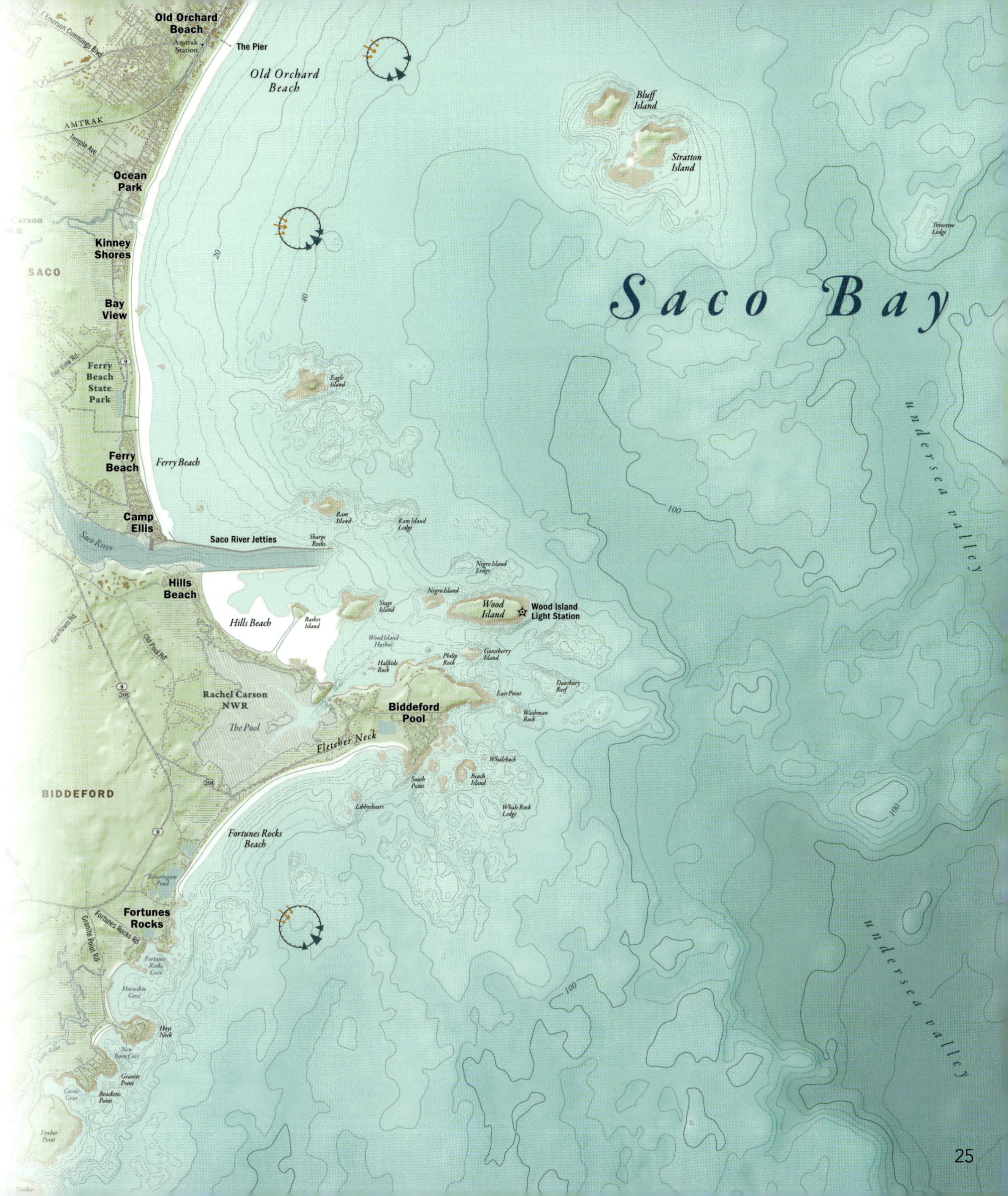

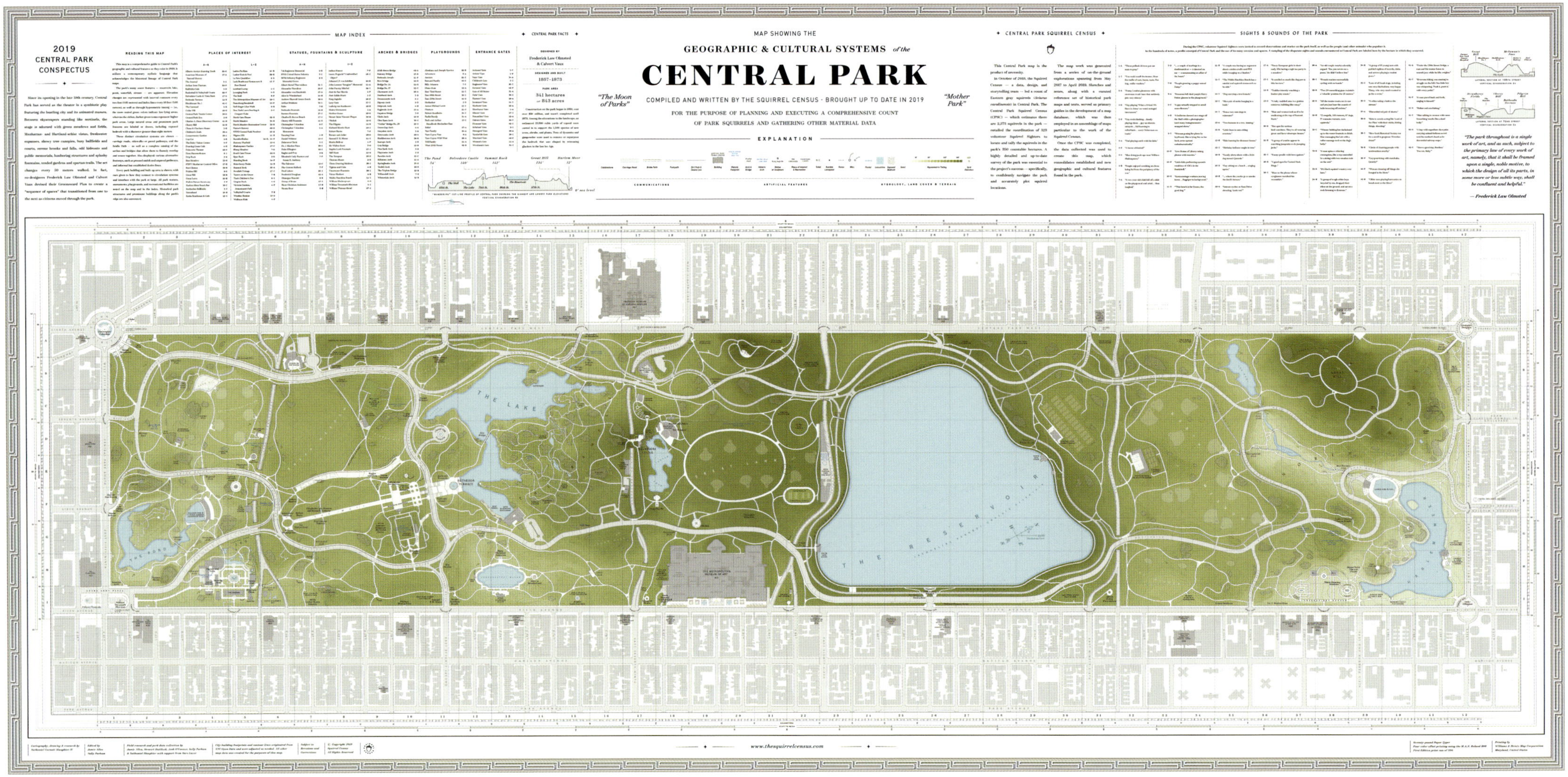

The Central Park Squirrel Census

Nathaniel Slaughter

In October 2018, the first official Central Park Squirrel Census was conducted in New York City. Throughout the project, from planning and data gathering to analysis and visualization, maps were a critical and essential tool. The Squirrel Census team surveyed and mapped Central Park, producing one of the most detailed map databases of the park of the past eighty years. This database provided the structure from which two significant maps were generated, serving as the foundation for the *Central Park Squirrel Census 2019 Report*.

The "terrestrial" map presents the geographic and cultural systems of the park. Terrain and geologic representation create a foundation for the map on which three primary communication systems—carriage roads, the Bridle Path, and footpaths—are placed. This circulatory network is interwoven through a complete catalog of arches and bridges, and connects the park's built-up areas—buildings, plazas, and hardscape gathering spots. Space

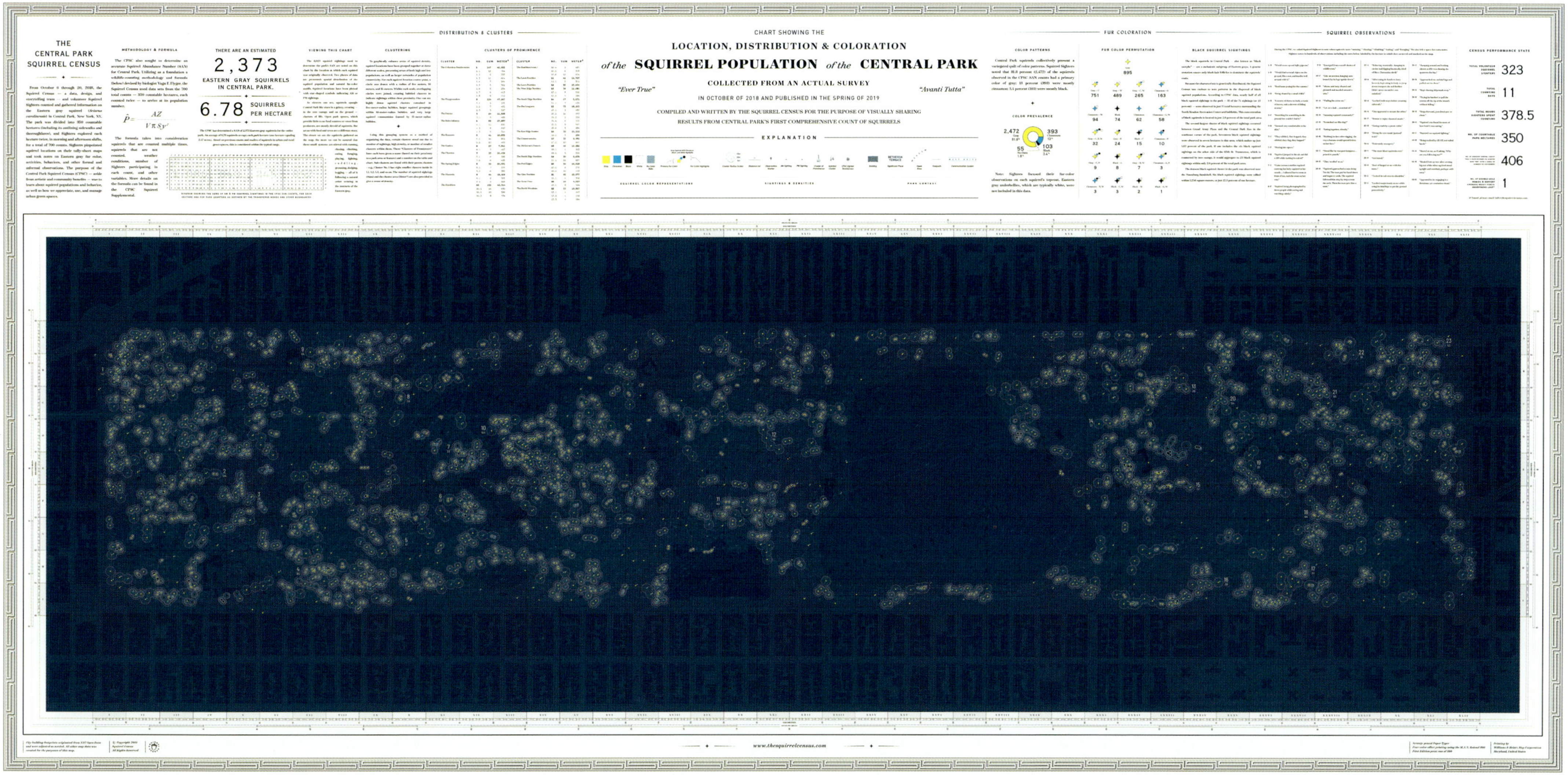

above the map is dedicated to detailed information about the park's content—both to encapsulate what was experienced in Central Park during October 2018 through census volunteers' park observations, and to encourage future park exploration through a comprehensive park index.

Whereas the park map is one of the land and the human experience, the squirrel map is one of the sky and the squirrel galaxy. This "celestial" map charts the location, distribution, and coloration of Central Park squirrels. A total of 3,023 squirrel sightings are shown with starlike symbols, and designated hues represent both primary and highlight fur colors. The map's color palette utilizes a primary background color; from this fulcrum, referential park geometries are presented in darker colors, while lighter colors are allocated to census data.

Spatial proximity of sighting points is shown through a three-tier series of clustering, the smallest of which emphasizes highly dense areas of sightings, while the largest tier shows sprawling-but-connected networks of squirrel activity. Twenty-three "Clusters of Prominence" have been identified, labeled, and referenced in a table, with each cluster entry showing the number of sightings and location. The clusters also have been given "team names" from a fictional Central Park squirrel softball league. In addition, select squirrel observations noted during the count are linked to the map by hectare ID.

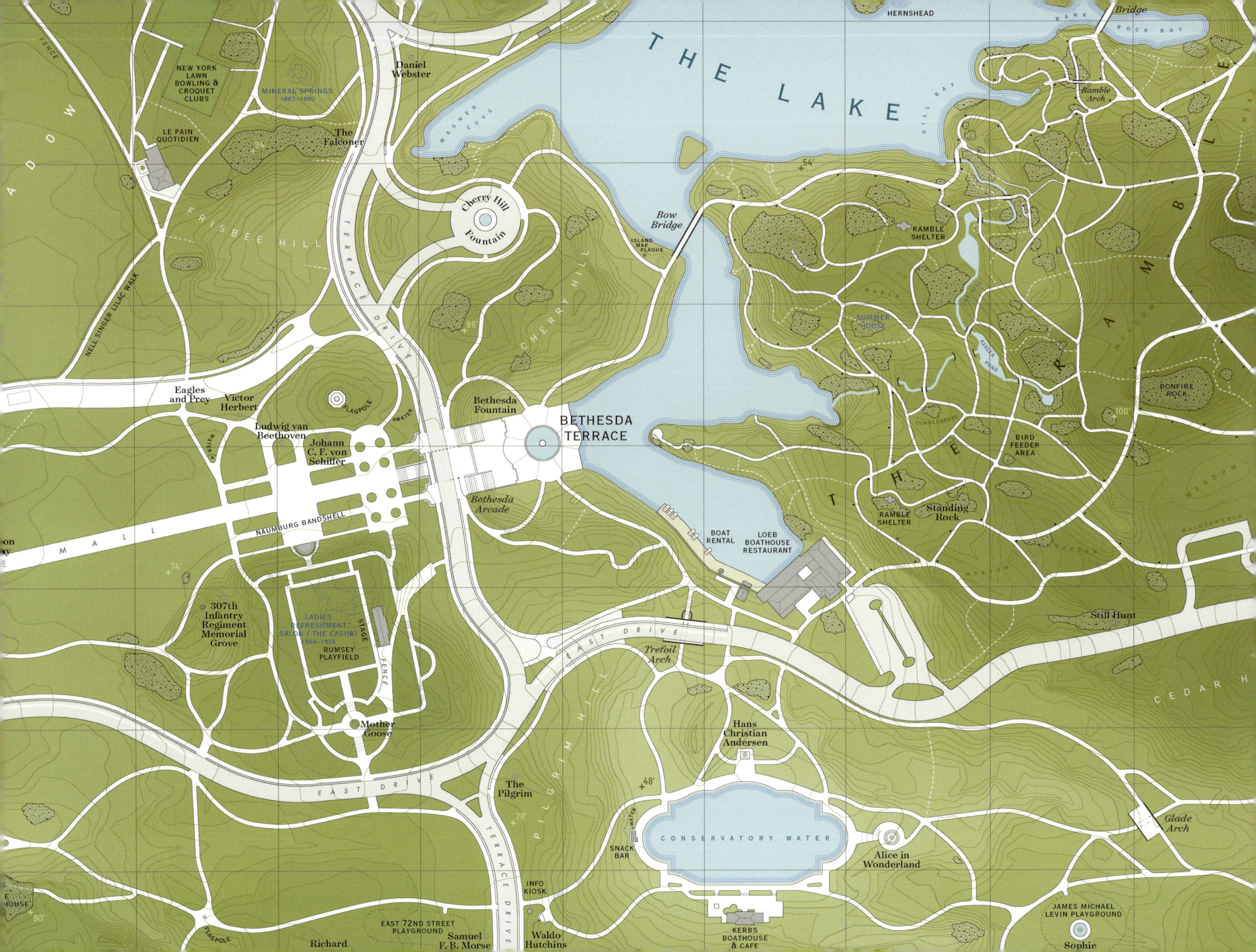

THE LAKE
HERNSHEAD
BANK
Bridge
ROCK BAY
GILL BAY
WAGNER COVE
Ramble Arch
Daniel Webster
NEW YORK LAWN BOWLING & CROQUET CLUBS
MINERAL SPRINGS 1867–1960
LE PAIN QUOTIDIEN
The Falconer
94'
54'
Cherry Hill Fountain
FRISBEE HILL
TERRACE DRIVE
Bow Bridge
ISLAND MAP PLAQUE
RAMBLE SHELTER
SUMMER HOUSE
AZALEA POND
BONFIRE ROCK
100'
BIRD FEEDER AREA
THE RAMBLE
MEADOW
NELL SINGER LILAC WALK
CHERRY HILL
86'
Eagles and Prey
Victor Herbert
FLAGPOLE
WATER
Bethesda Fountain
BETHESDA TERRACE
Ludwig van Beethoven
Johann C. F. von Schiller
Bethesda Arcade
NAUMBURG BANDSHELL
MALL
74'
BOAT RENTAL
LOEB BOATHOUSE RESTAURANT
RAMBLE SHELTER
Standing Rock
Still Hunt
307th Infantry Regiment Memorial Grove
LADIES REFRESHMENT SALON / THE CASINO 1864–1936
STAGE
RUMSEY PLAYFIELD
FENCE
EAST DRIVE
Trefoil Arch
CEDAR HILL
Mother Goose
Hans Christian Andersen
PILGRIM HILL
48'
The Pilgrim
78'
CONSERVATORY WATER
SNACK BAR
Alice in Wonderland
Glade Arch
INFO KIOSK
EAST 72ND STREET PLAYGROUND
Samuel F. B. Morse
Waldo Hutchins
KERBS BOATHOUSE & CAFE
JAMES MICHAEL LEVIN PLAYGROUND
Sophie
Richard
80'

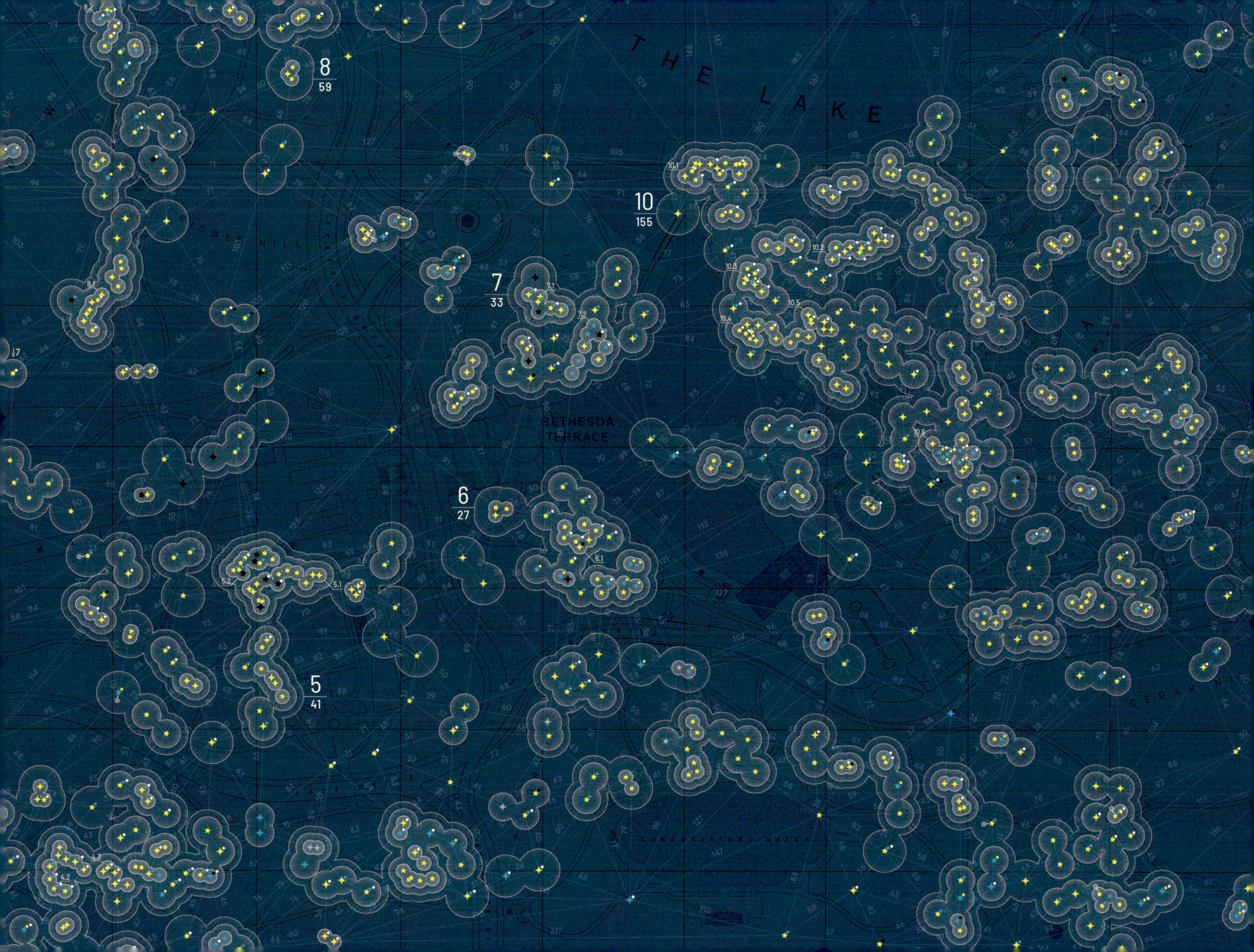
THE LAKE
8
59
10
155
7
33
BETHESDA
TERRACE
6
27
5
41

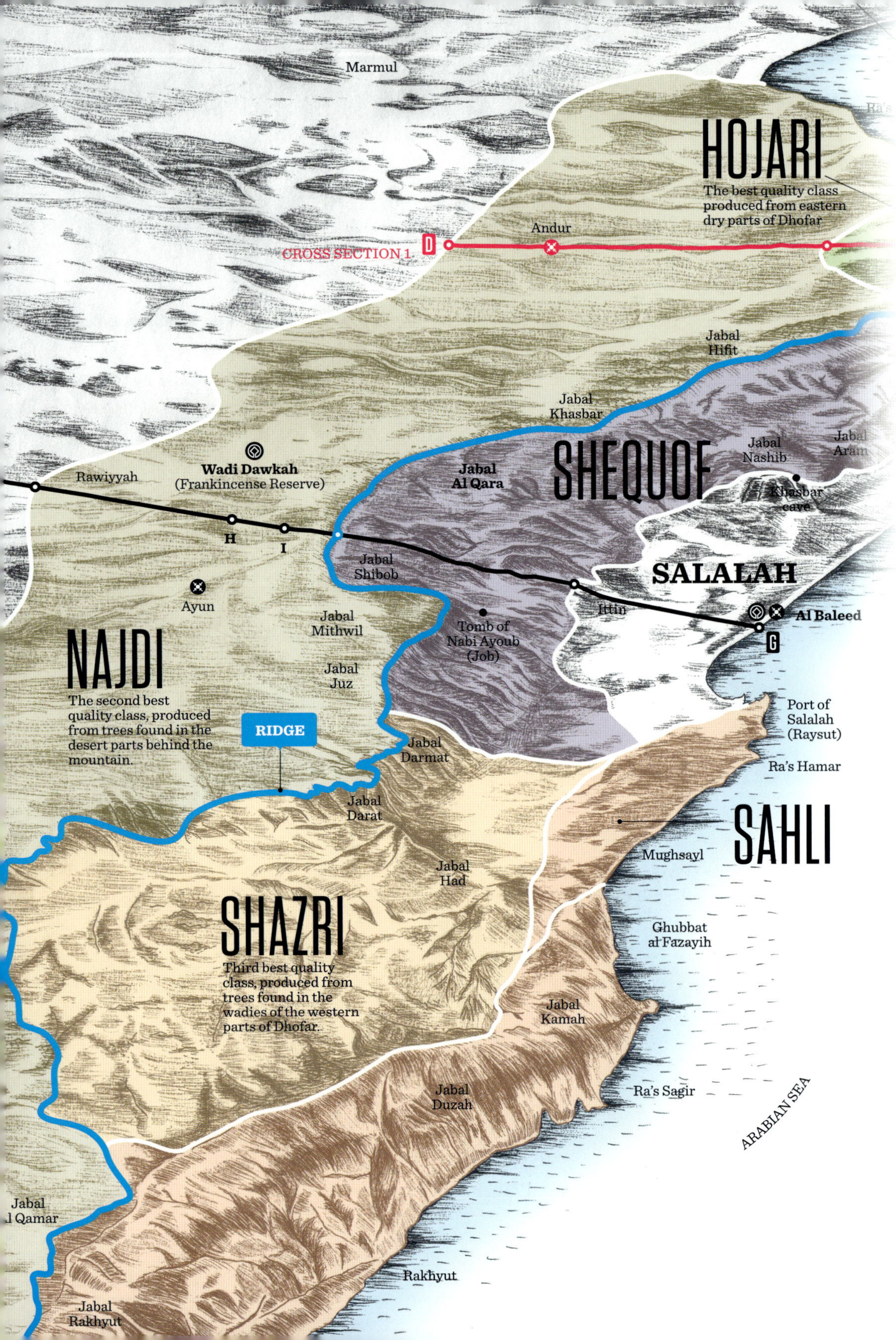

OMAN: THE LAND OF FRANKINCENSE

The Zubair Corporation

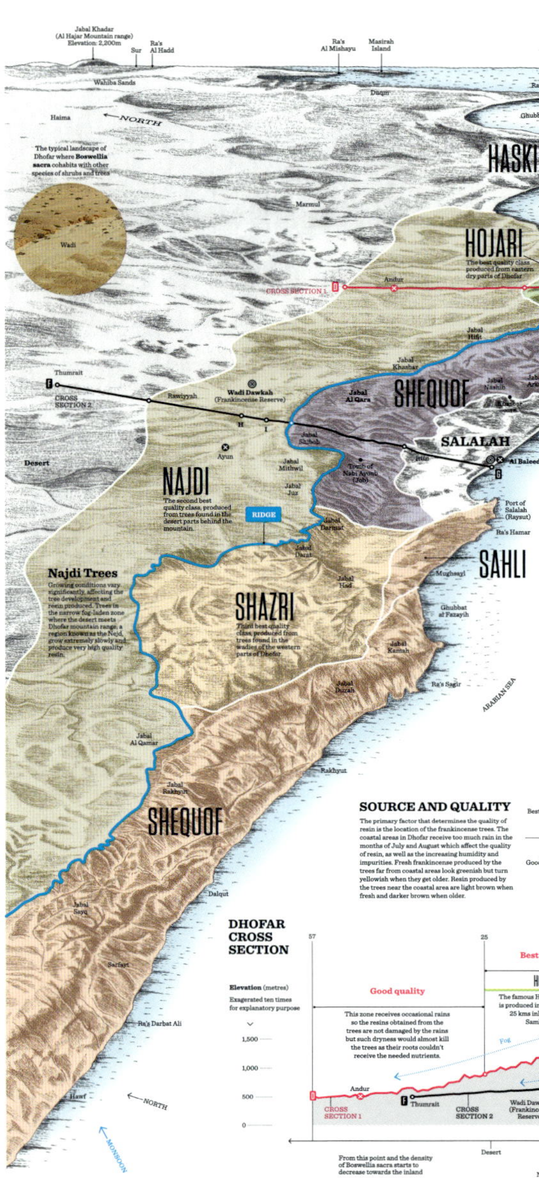

THE BLESSED DHOFAR REGION

Dhofar, located at the southern part of Oman, prides itself as the home of the world's best quality frankincense resin known as the *Royal Hojari*. This is due to the fact that Dhofar's perfectly situated geographic location produces quite unique weather. To understand this map, start from the ridge of Dhofari mountain chain (blue line) which divides Dhofar into two main domains: the coastal plain which annually receives monsoon rains (or *khareef*), and the plateau, a drier tract which produces the best frankincense.

Jabal = Mountain **Wadi** = Dried up riverbed Ancient frankincense collection point Unesco site

SOURCE AND QUALITY

The primary factor that determines the quality of resin is the location of the frankincense trees. The coastal areas in Dhofar receive too much rain in the months of July and August which affect the quality of resin, as well as the increasing humidity and impurities. Fresh frankincense produced by the trees far from coastal areas look greenish but turn yellowish when they get older. Resin produced by the trees near the coastal area are light brown when fresh and darker brown when older.

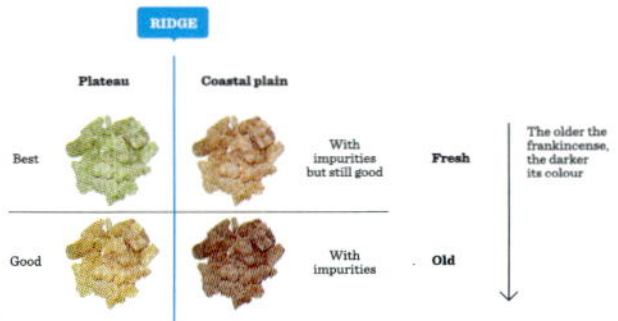

DHOFAR CROSS SECTION

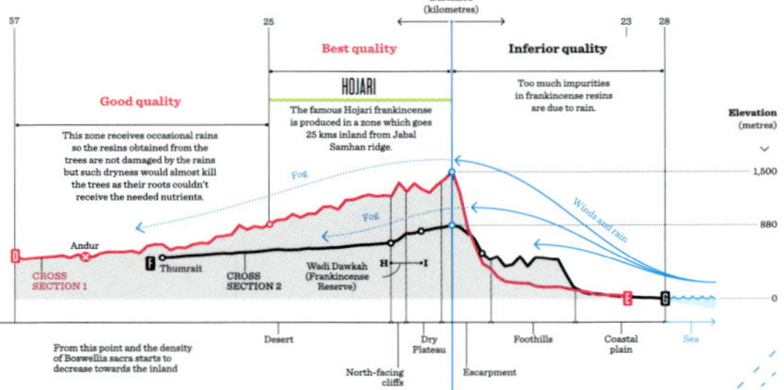

SIZE AND SHAPE OF THE TREE

Plant species have certain range and characteristic. They are greatly dependent on the climate, the availability of water, quality of sunlight, temperature and soil. Thus, the Boswellia sacra in Oman may vary in many ways:

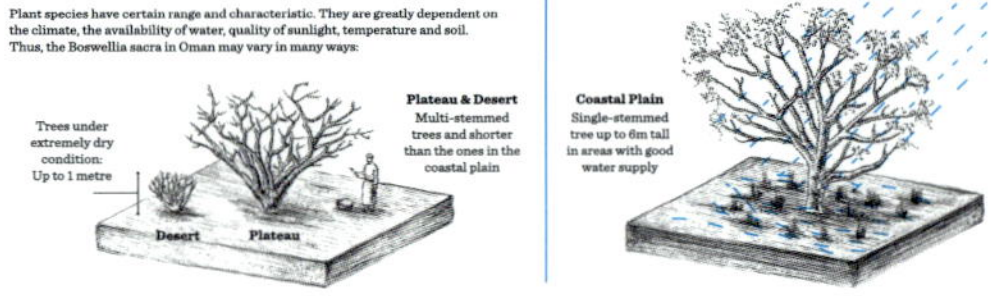

VANTAGE POINT

The map above is seen from the coast of Yemen facing the Sultanate and the location where frankincense trees abundantly grow spanned within the area of the page. Note however that in this perspective the scale is a bit skewed due to the roundness of the Earth's surface.

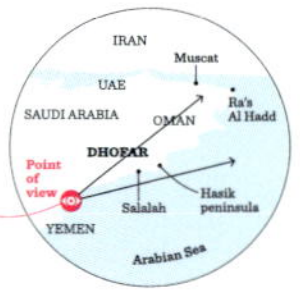

Sources: *Burseraceae of Dhofar* by Ahmed Mohammed Mustahil Al Kathiri

THE FRANKINCENSE TRAIL

A network of trade routes had been established between kingdoms for many centuries, carrying goods such as salt, wheat, wine, weapons, dates, and animal skins from one fortified settlement to another. These paths were the basis of the frankincense trade routes between the kingdoms of southern Arabia, which formed the starting point of the route towards the north. Frankincense was a sought after commodity throughout Egypt, Rome, Greece, Persia, Assyria and Babylon and was one of the most prized substances of the ancient civilized world, valued as highly as gold and the gift of kings. For centuries, incense trade routes from Arabia Felix to the Mediterranean where specifically made via camel caravans, although, this was not the only route. Several possible routes varied over time. From the Dhofar region where the frankincense trees were grown, our Arabian ancestors had to make several long journeys trough some hostile territories.

Land of Frankincense

Wadi Dawkah, Shisr, Khor Rori and Al Baleed are the four components known collectively as the Land of Frankincense declared in the UNESCO World Heritage list. They dramatically illustrate the trade in frankincense that flourished in this region for many centuries and constitute outstanding testimony to the civilizations in south Arabia since the Neolithic age.

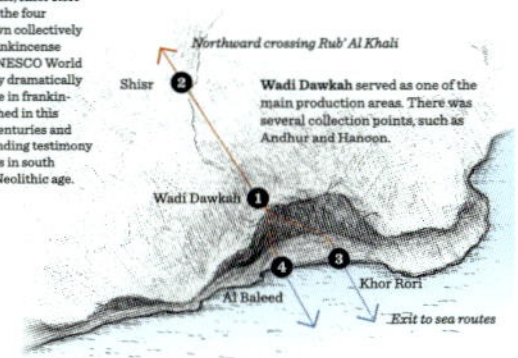

1 Wadi Dawkah

This natural park represents a region where the frankincense tree grows. Estimated by five sqare kilometers.

2 Shisr

A caravansary and an oasis where camel trains entering the desert of Rub Al Khali (The Empty Quarter) stopped for water.

3 Khor Rori

Hadhramaut Kingdom established Sumhuram in order to control main routes to the Red Sea, Arabian Gulf and Indian Ocean.

4 Al Baleed

Al Baleed was a trade hub and a commercial centre with ships leaving from Salalah to carry cargoes of frankincense to China and Rome. Chinese traveller Jan Jokao wrote that frankincense was one of the main commodities produced by Al Baleed. The Chinese city of Guangzhou imported about 174 kg of resins from this town.

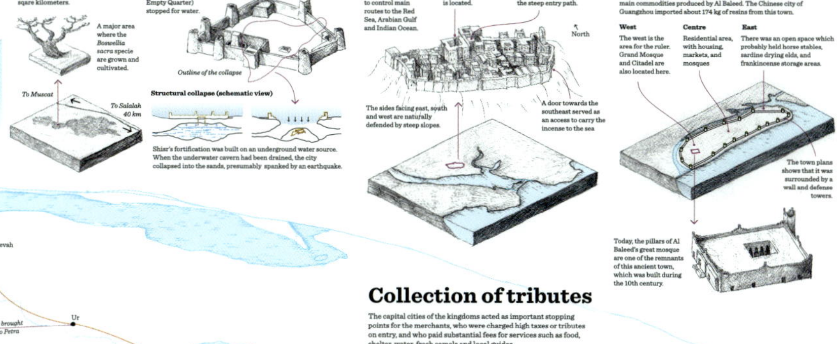

Inland routes

Route 1
The oldest land route. Starts from Ain Humran (Dhofar) and travelling north to Shisr, then across the Empty Quarter to Jabrin, Gerrah, and Mesopotamia.

Route 2
From Sumhuram west (inland) to Shabwa, then north through Dedan and Petra until it reaches Gaza and Mediterranean. Also a connection with **Route 3**, and north to Jabrin linking with **Route 1**.

Route 2A
From Petra, the route branched in five directions, most significant was the one heading to Alexandria (or Gaza, Jerusalem, Jerash) then to Damascus.

Route 2B
From Hegra, routes went to west via Myos Hormos and Caenopolis, crossing Red Sea, then onwards to Alexandria.

Alternative routes connecting the other main inland routes.

ANCIENT TRADE ROUTES

The three ancient routes converged in Palmyra and Petra. Frankincense route was the oldest.

Collection of tributes

The capital cities of the kingdoms acted as important stopping points for the merchants, who were charged high taxes or tributes on entry, and who paid substantial fees for services such as food, shelter, water, fresh camels and local guides.

Shisr
At Wubar a tenth of the cargo would be demanded.

Shabwa
Caravans had to pay 10% of its loads.

Petra
Traders were to pay a quarter of the goods upon entering Nabataean kingdom.

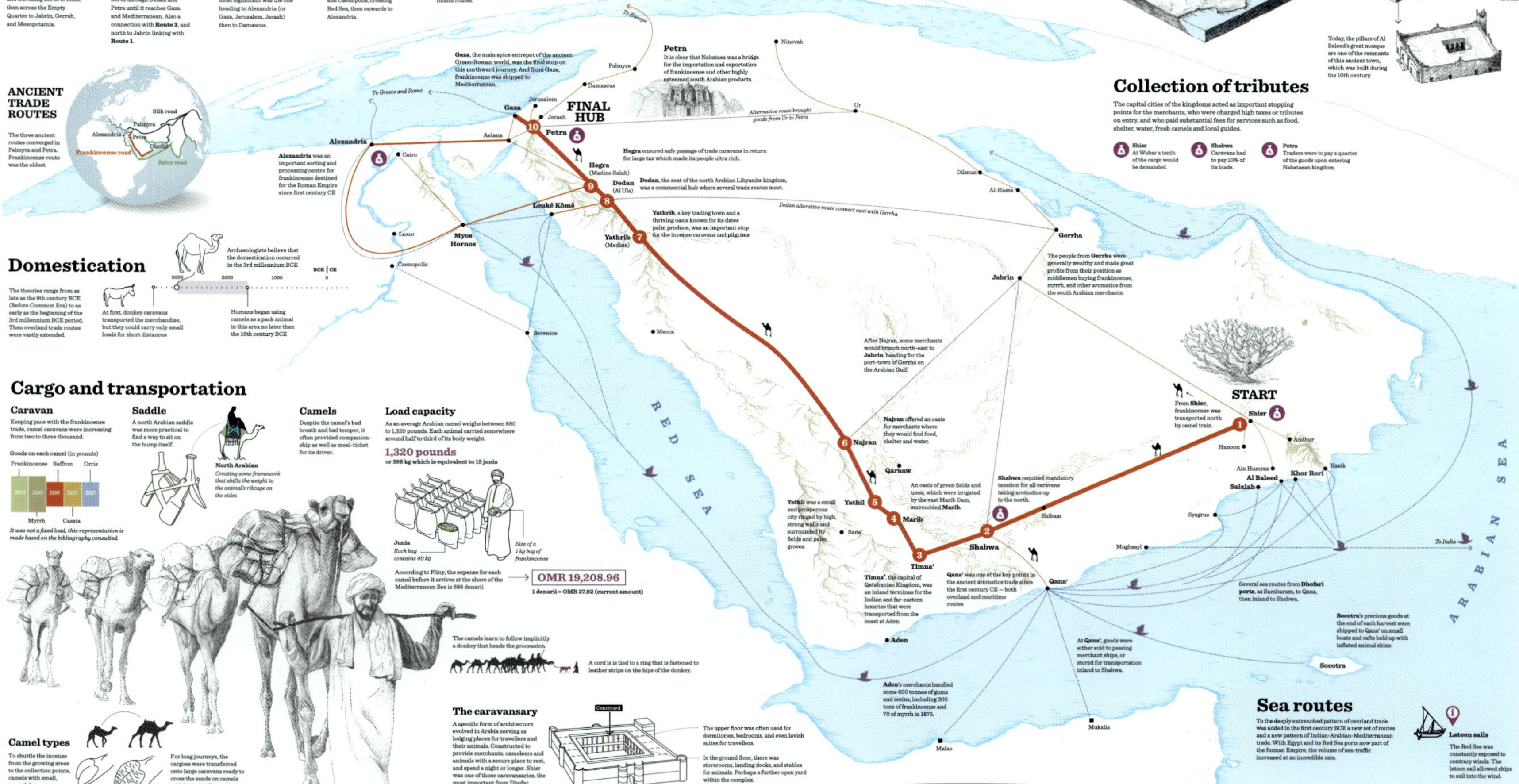

Domestication

Archaeologists believe that the domestication occurred in the 3rd millennium BCE

3000 | 2000 | 1000 | BCE | CE 0

The theories range from as late as the 9th century BCE (Before Common Era) to as early as the beginning of the 3rd millennium BCE period. Then overland trade routes were vastly extended.

At first, donkey caravans transported the merchandise, but they could carry only small loads for short distances

Humans began using camels as a pack animal in this area no later than the 16th century BCE

Cargo and transportation

Caravan

Keeping pace with the frankincense trade, camel caravans were increasing from two to three thousand.

Goods on each camel (in pounds)

Frankincense	Myrrh	Saffron	Cassia	Orris
300	300	200	200	200

It was not a fixed load, this representation is made based on the bibliography consulted.

Saddle

A north Arabian saddle was more practical to find a way to sit on the hump itself.

North Arabian
Creating some framework that shifts the weight to the animal's ribcage on the sides.

Camels

Despite the camel's bad breath and bad temper, it often provided companionship as well as meal-ticket for its driver.

Load capacity

As an average Arabian camel weighs between 880 to 1,320 pounds. Each animal carried somewhere around half to third of its body weight.

1,320 pounds
or 598 kg which is equivalent to 15 junia

Junia
Each bag contains 40 kg

Size of a 1 kg bag of frankincense

According to Pliny, the expense for each camel before it arrives at the shore of the Mediterranean Sea is 688 denarii → **OMR 19,208.96**

1 denarii = OMR 27.92 (current amount)

The camels learn to follow implicitly a donkey that heads the procession.

A cord is is tied to a ring that is fastened to leather strips on the hips of the donkey.

Camel types

To shuttle the incense from the growing areas to the collection points, camels with small, smooth hooves were employed.

For long journeys, the cargoes were transferred onto large caravans ready to cross the sands on camels with bigger, calloused but soft-soled hooves.

The caravansary

A specific form of architecture evolved in Arabia serving as lodging places for travellers and their animals. Constructed to provide merchants, cameleers and animals with a secure place to rest, and spend a night or longer. Shisr was one of those caravansaries, the most important from Dhofar northward.

Courtyard

The upper floor was often used for dormitories, bedrooms, and even lavish suites for travellers.

In the ground floor, there was storerooms, landing docks, and stables for animals. Perhaps a further open yard within the complex.

Sea routes

To the deeply entrenched pattern of overland trade was added in the first century BCE a new set of routes and a new pattern of Indian-Arabian-Mediterranean trade. With Egypt and its Red Sea ports now part of the Roman Empire, the volume of sea-traffic increased at an incredible rate.

Lateen sails

The Red Sea was constantly exposed to contrary winds. The lateen sail allowed ships to sail into the wind.

Route elevation and stopovers

According to Pliny the Elder, the journey consisted of 65 stages divided by halts and stopovers for the camels, and places for merchants to pay for services.

Shabwa was the mandatory point of taxation for all the caravans. Every merchant would have had to enter by law, and on pain of death if someone tried to pass another way long, during his journey.

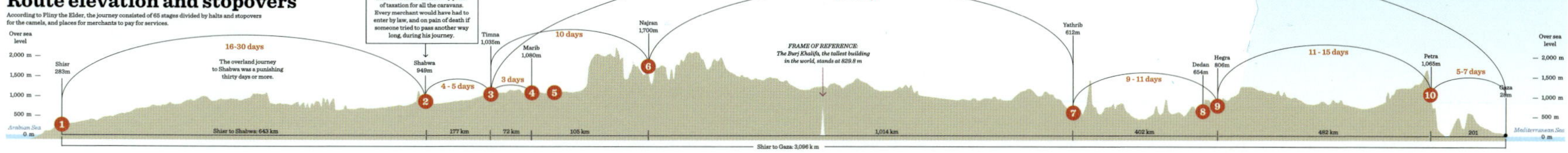

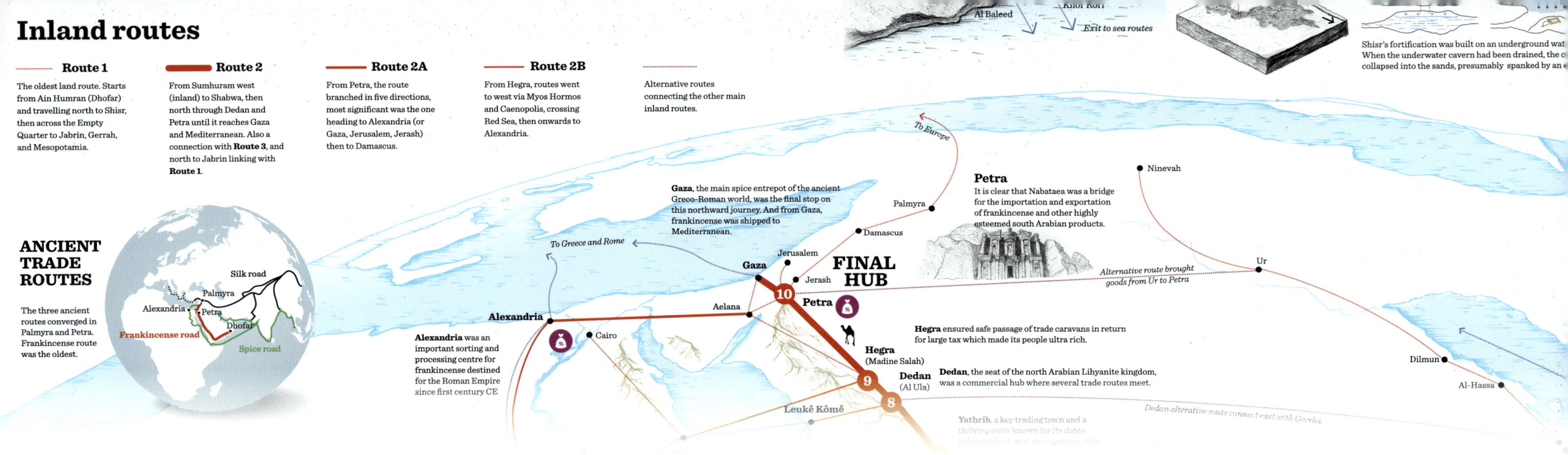

« The Blessed Dhofar Region/The Frankincense Trail

Adonis Durado, Antonio Farach, Jorge Cortés, Isidore Vic Carloman, Winie Ariany, Erik Tuban, Muthita Torteeka, and Ismael Sandiego — Times of Oman

Oman, the Land of Frankincense

The solidified resin of the *Boswellia sacra* tree, known as "frankincense," has been extracted and traded for millennia by the people of Oman. It's a matter of high pride. To celebrate the Oman National Day, our team produced an infographic about frankincense that included two maps.

The Blessed Dhofar Region

This map shows the areas where the different grades of frankincense are produced and how the climate affects the area. Given that both climate and terrain-shape are the crucial determinants of frankincense quality, I knew that my main storytelling "devices" needed to be a panoramic map and cross-section visuals. These were both essential to show how the steep mountains block the monsoon rains, allowing only mist to pass toward the best lands for frankincense.

The best frankincense comes from arid regions, and I wanted to highlight that, so we used color—a blue line—to work as the main visual narrative driver. The line helps to highlight the separation between the areas that only receive mist, versus the lands that receive rain.

For these, we also used the U.S. Geological Survey *Atlas of Oblique Maps* as design inspiration.

The Frankincense Trail

Frankincense was in high demand by the Greeks, Egyptians, Persians, and other civilizations of the Far East, but its greatest boom occurred at the time of the Roman empire, when it was traded and used in large quantities and helped produce a trade route toward the Mediterranean.

This illustrated map shows the main routes that camel caravans and ships traveled to transport Omani incense. We maintained monochromatic textures in the map to let the colorized routes lead the eye into the areas where we wanted readers to focus.

We conducted extensive research, which included in-situ study at the remains of the original land and maritime ports from which the frankincense was shipped. We 3D-modeled and then traced these hubs using ballpoint pen on paper, in order to show the final relief rendering on the map.

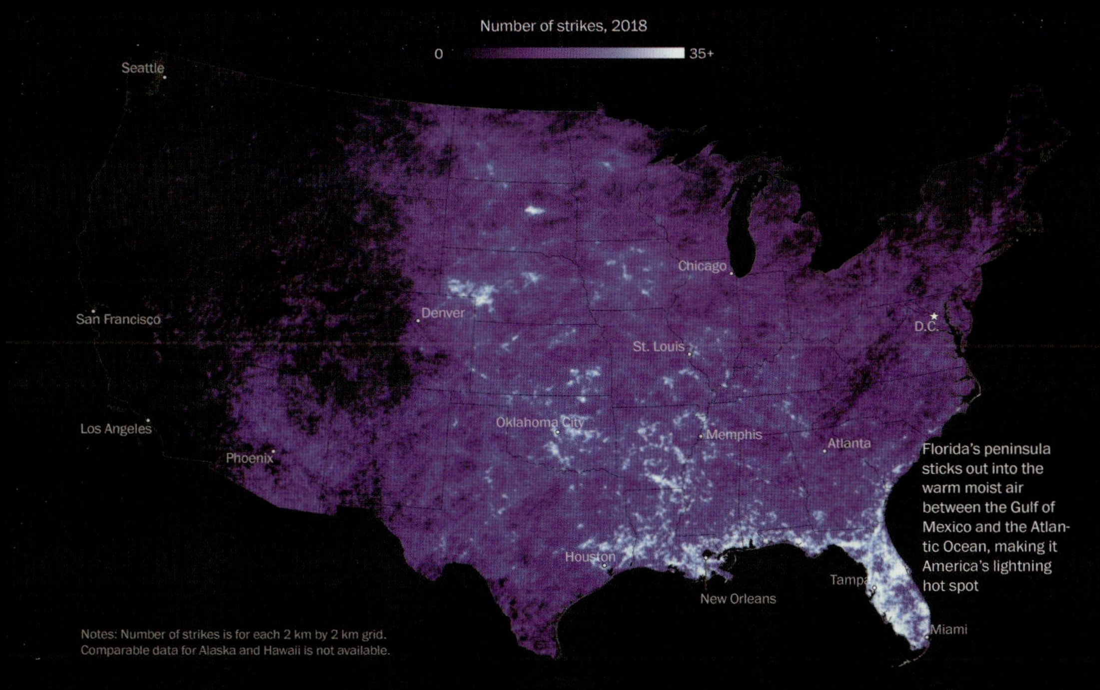

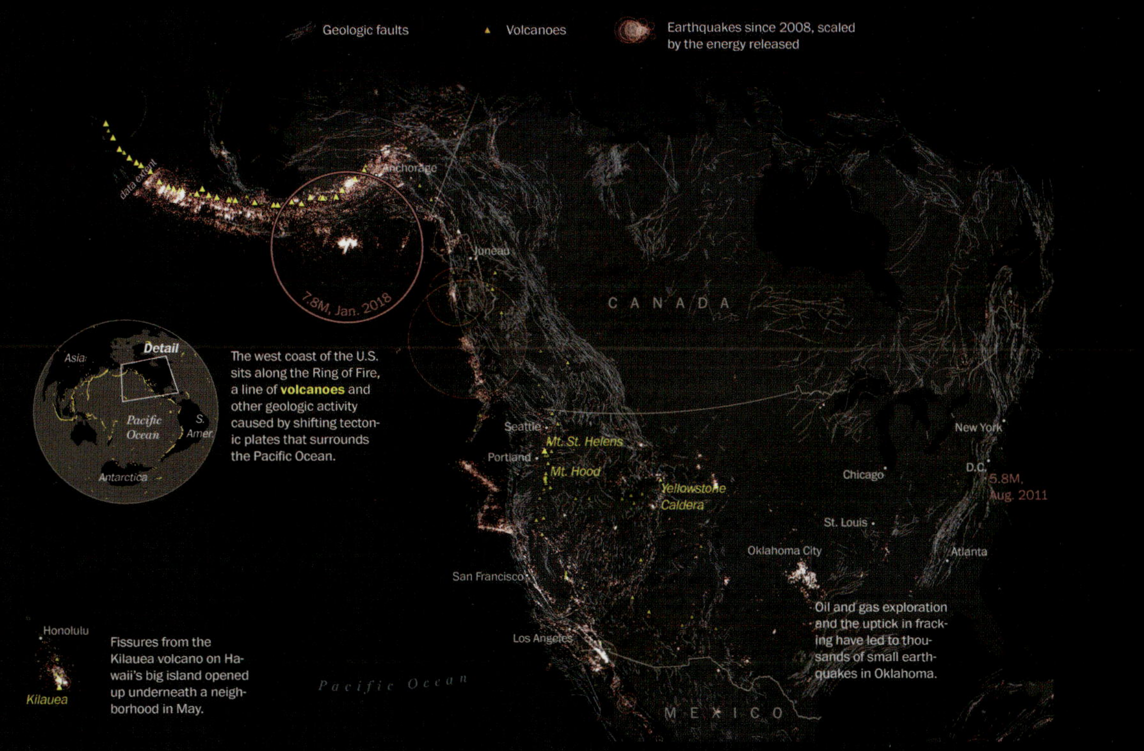

Mapping America's Wicked Weather and Deadly Disasters »

Tim Meko — The Washington Post

The fear of losing everything in a natural disaster sits in the back of our minds no matter where we live. The same diverse physical geography that gives us sunny beaches and crisp mountain air also generates devastating storms and wildfires. Climate change is only making things worse.

Data collection for these events has never been more consistent. Mapping the trends in recent years gives us an idea of where disasters have the tendency to strike. In 2018, it is estimated that natural disasters cost the nation almost $100 billion and took nearly 250 lives. It turns out there is nowhere in the United States that is particularly insulated from everything.

The kernel of an idea for this project had been in my mind for years. I sent my first data request to a scientist at NOAA in early 2017. I called the project the *Map of Doom*. After being sidetracked by months of vicious news cycles (many of them disaster-related), I finally scraped together enough free time and data to test my original thesis: that there is no place truly safe from everything.

The design of the maps was focused on the story—the confluence of risk and geography. I used vibrant colors set upon a dark background and semi-transparent layers in order to highlight the most frequent areas of impact for a given disaster. Simple annotations help explain interesting tidbits on the maps, but generally I refrained from too many labels, which can be distracting. State outlines suffice to help readers find their place within the map, but are so light that they fade to the background if you aren't looking for them. The same goes for a faint hillshade layer—the terrain is partly to blame for many of the phenomena in the maps, but it's not the star of the show.

Though dangerous, these events have found a way into our identities. Sports teams embrace their local disasters: the cyclones, hurricanes, heat, and avalanche. In many ways, there is pride in the resilience we feel from being constantly bombarded by catastrophes. What I find most uplifting is how they bring out the best in our communities.

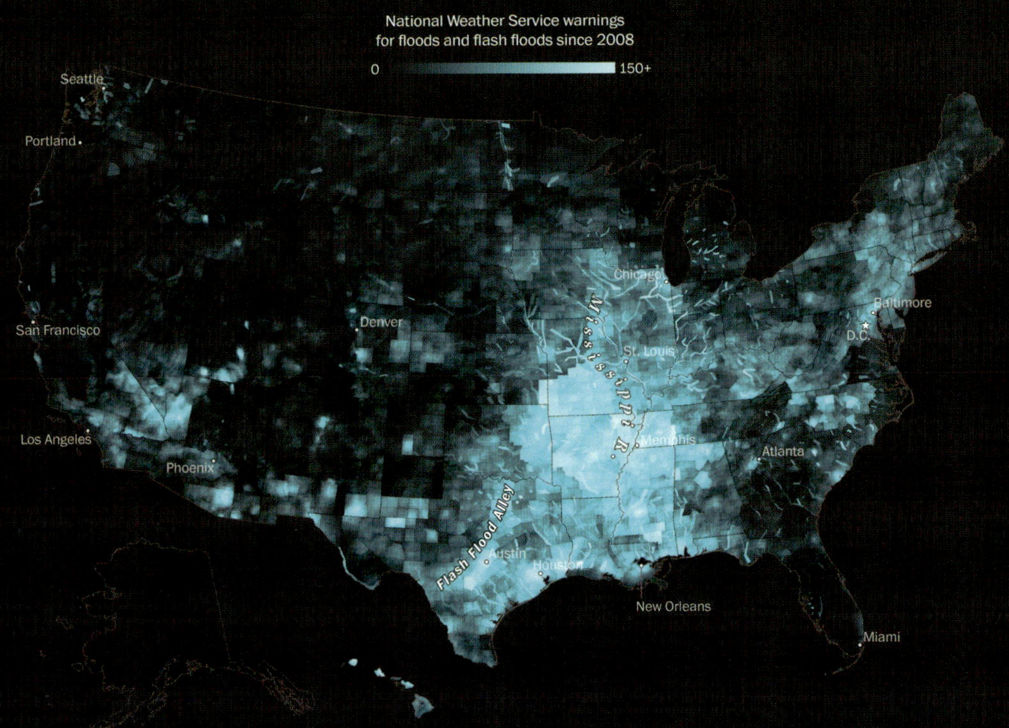

Note: Only includes areas where the NWS issued flood warnings for specific areas, not the entire county.

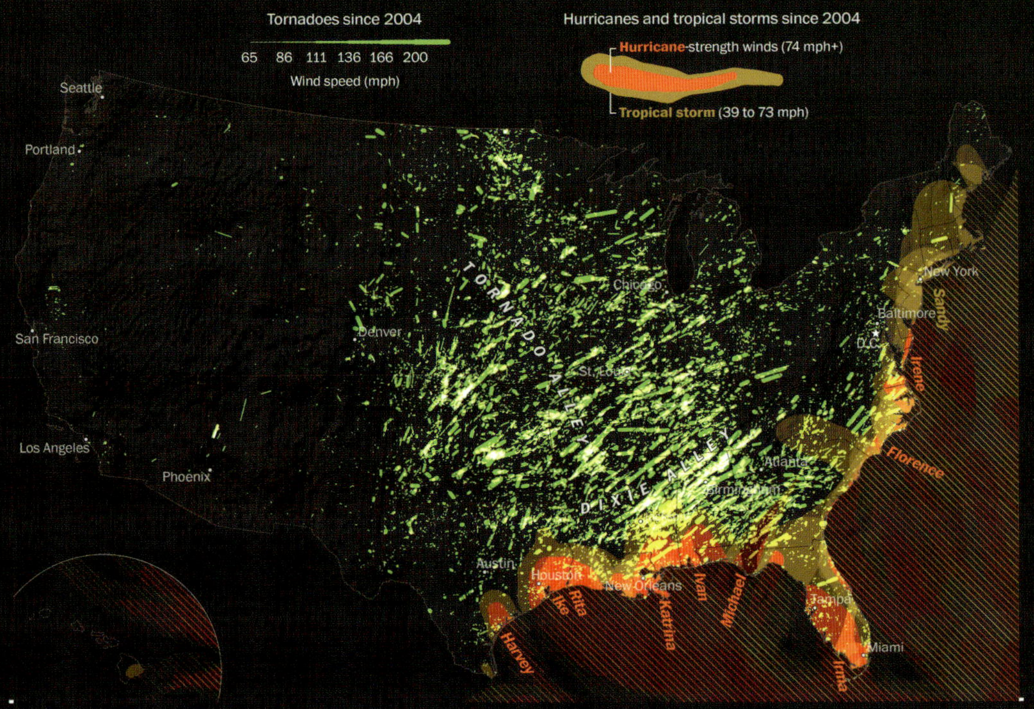

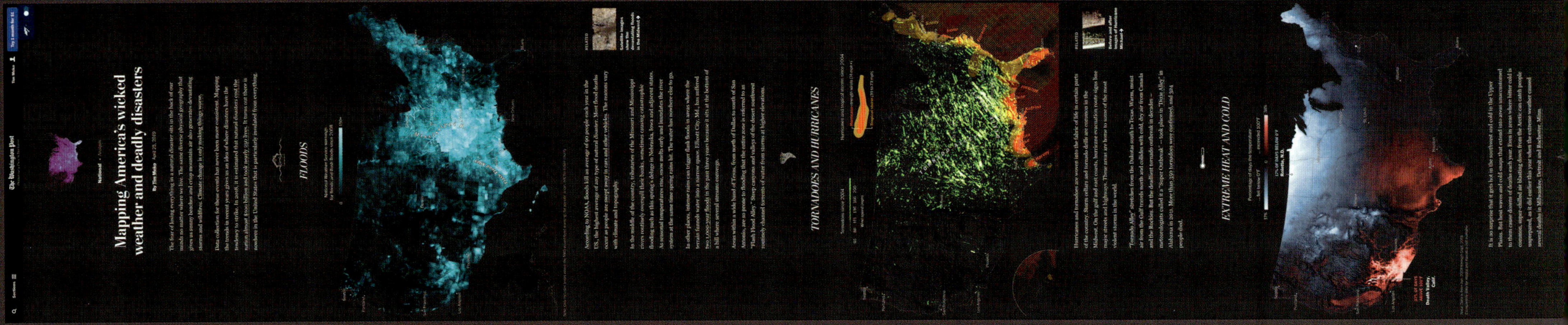

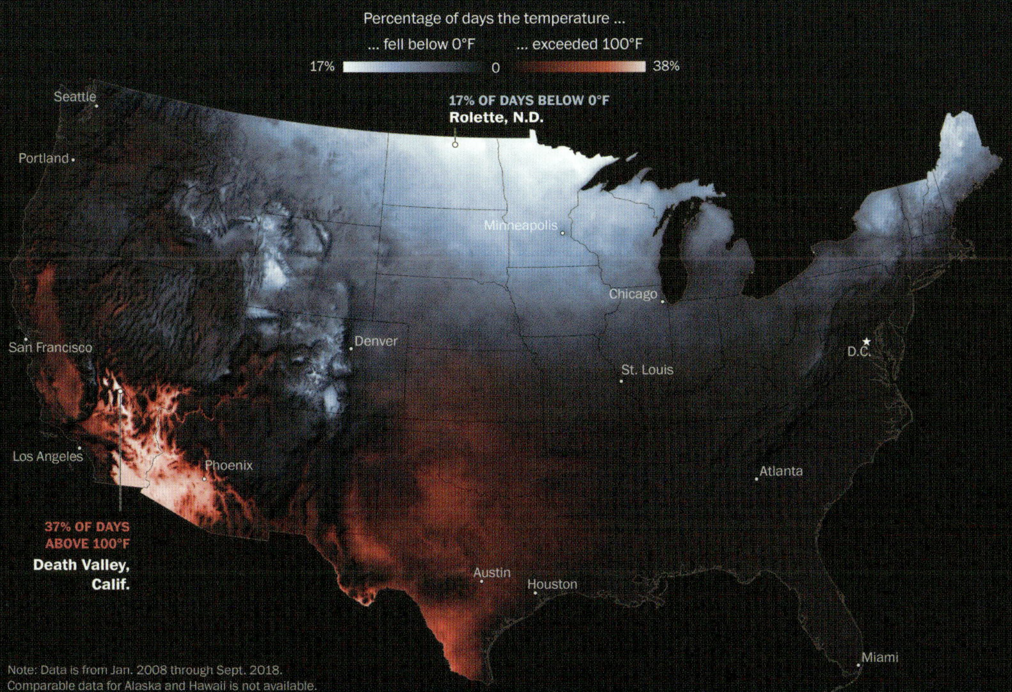

Note: Data is from Jan. 2008 through Sept. 2018.
Comparable data for Alaska and Hawaii is not available.

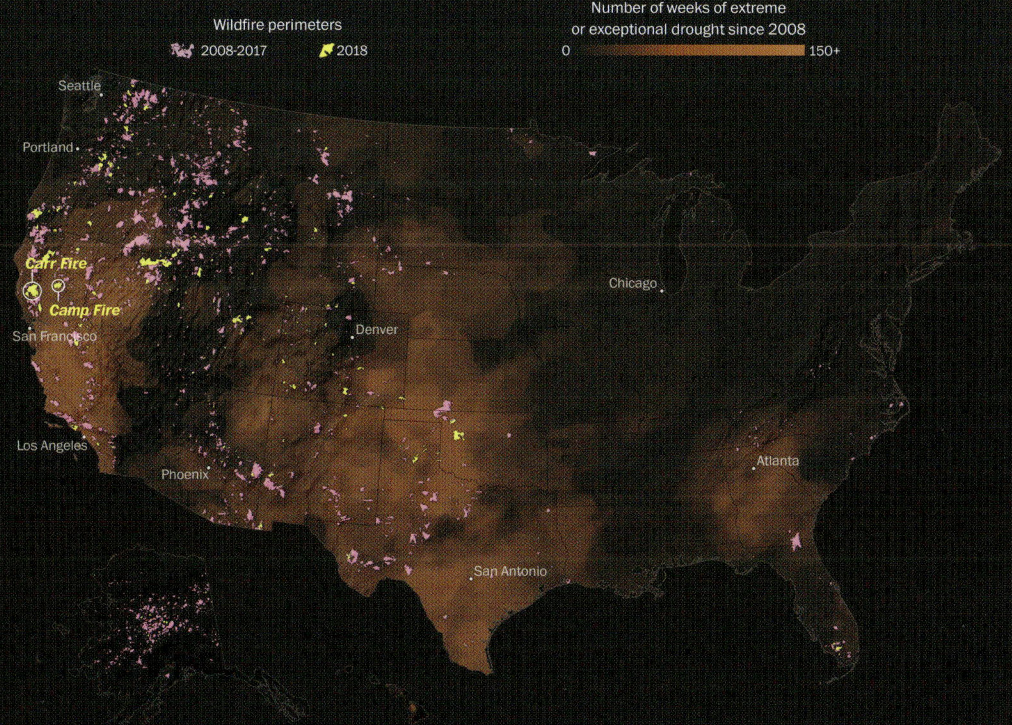

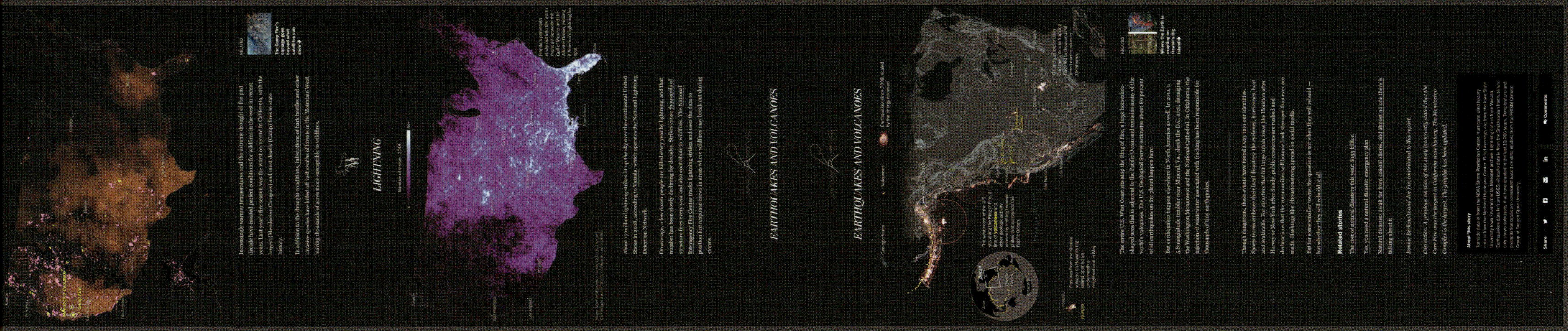

A CHRONICLE OF CRATERS

In the last century, humans have spread from the Cassaneva crater, building settlements and cities along the rivers and in the valleys of southwest Batīvan. Although they mainly hunt with arrows and spears, and practice subsistence agriculture, their presence has pushed the Red Mayaleta from the plains where they once ruled, to the forests where they have been forced to adapt to a new landscape.

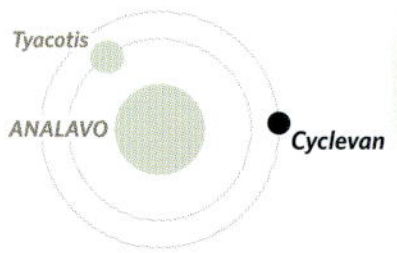

ANALAVO SYSTEM

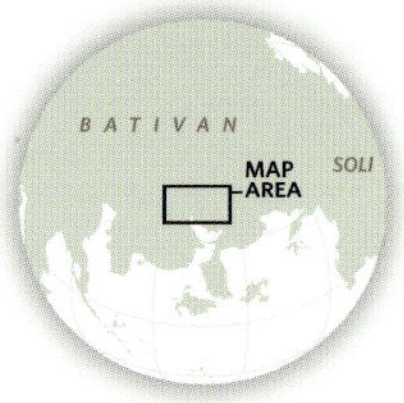

CYCLEVAN MOON

RED MAYALETA

Unlike the other species of mayaleta, the Red Mayaleta relies on a mixed diet of ground flora and small mammals. Traditionally living on the fringe of forests, they make use of the open prairie to graze and the forest to retreat to when raising their young.

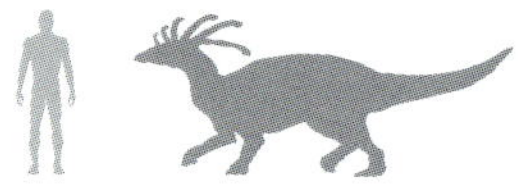

AZCOMBI MAYALETA

Though rarer, the Azcombi Mayaleta is much larger and more resilient than other species of its kind. Only eating several times a week, these creatures roam the high deserts.

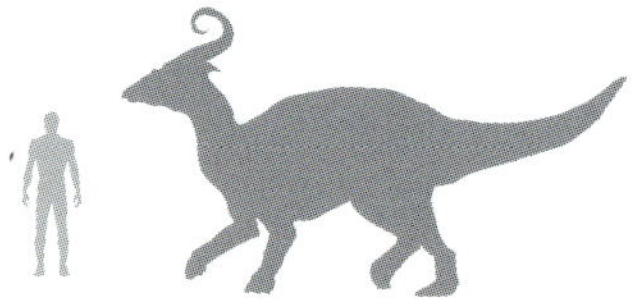

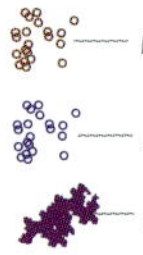

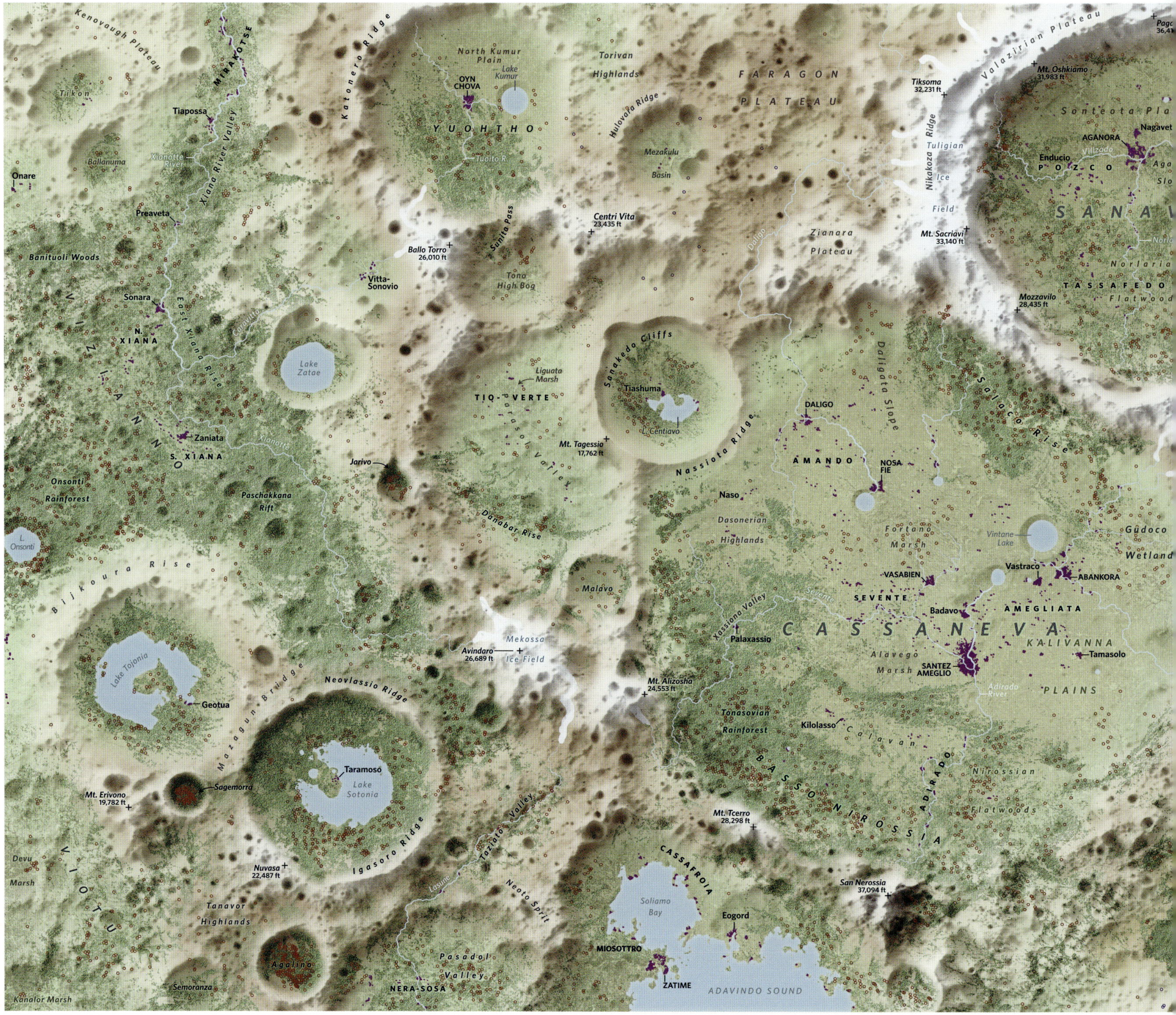

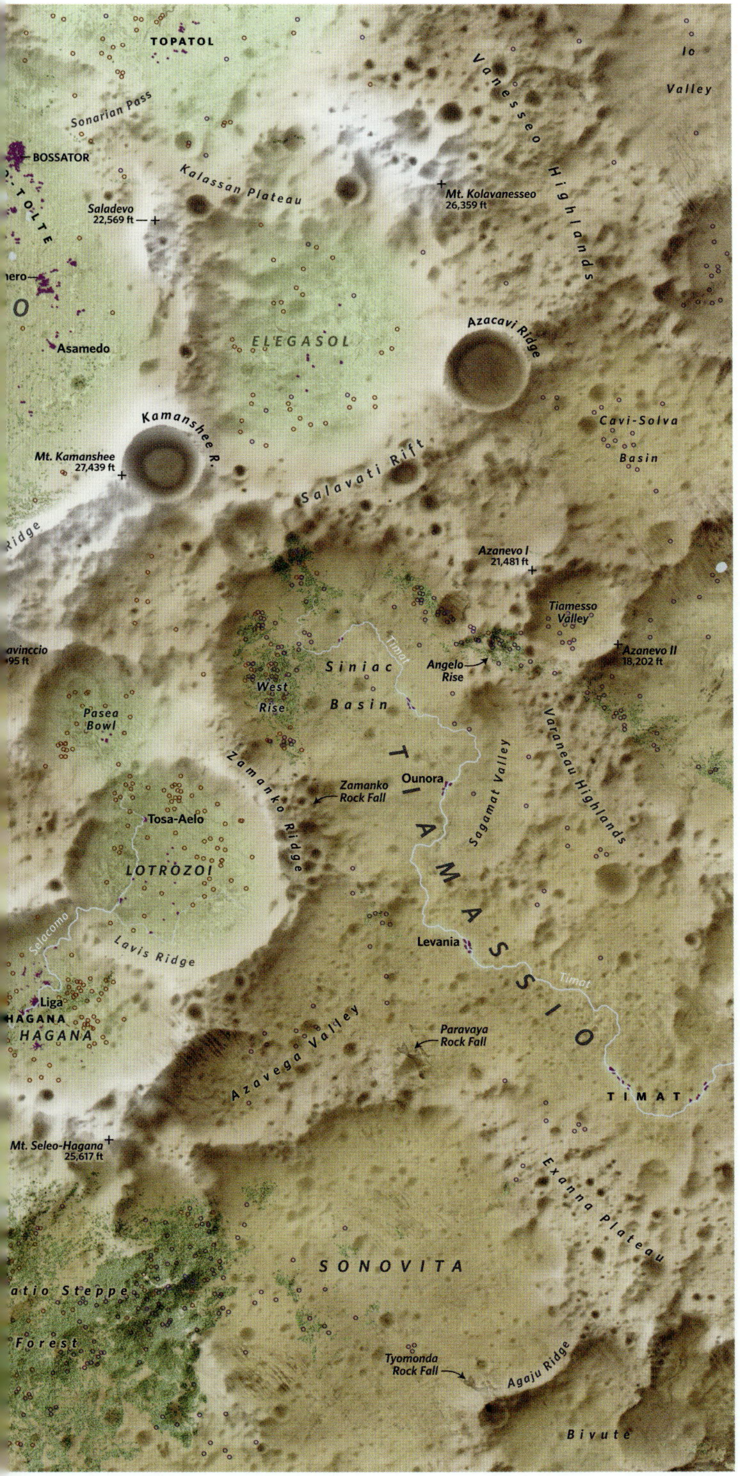

« A Chronicle of Craters

Soren Walljasper

I've always been interested in the way that Planet Earth can house so many types of landscape. I've spent hours looking at aerial imagery, picking apart the syntax of physical geography. The lush central lowlands of North America fade into the dry Great Plains as the elevation slowly rises. The Cascades act as a rain shadow, separating dense forest from a barren desert. These nuances intrigued me. During the fiftieth anniversary of the moon landing in July 2019, when everyone was looking at the craters above us, I couldn't help but think about how Earth-like terrain would play out over the ridges and bowls of a lunar surface. So, one Sunday during a #creativeCarto meeting (a monthly event where cartographers gather online via Twitter or Slack to learn and create), I began the simple task of creating my own fictional planet.

My primary goal was making this imaginary landscape look realistic, and the key to that was detail. I had to mimic the soft textures and precise relief that create an authoritative feeling, so I looked at styles from trusted sources like the National Park Service, *The Washington Post*, and the National Geographic Society. To achieve this believability, I had to pull data from the real world. I started with an elevation model from the Earth's moon. From there I blended in land cover, mixing and contorting different aerial imagery and datasets from all over the globe to mimic a realistic landscape. I filled low-lying craters with green imagery from the Midwest, layered high plateaus with texture from the Sahara Desert, and combined forest masks from all over the world to make something that had verisimilitude.

Once I had a believable landscape, it was time to fill out my story. I envisioned how humans and nature would interact across the ridges and plains of this fictitious land. I thought about the agriculture and trade that must happen there—where the cruxes of flows would best support cities and towns, and how those settlements would affect surrounding wildlife. As I added more complexity to the map, I found myself needing to lighten the terrain and assign the "data" presented on top of it bright, unnatural colors to step off the background. At the end of the day, I had a carto-graphic novel that told a story about a wild planet, far from Earth.

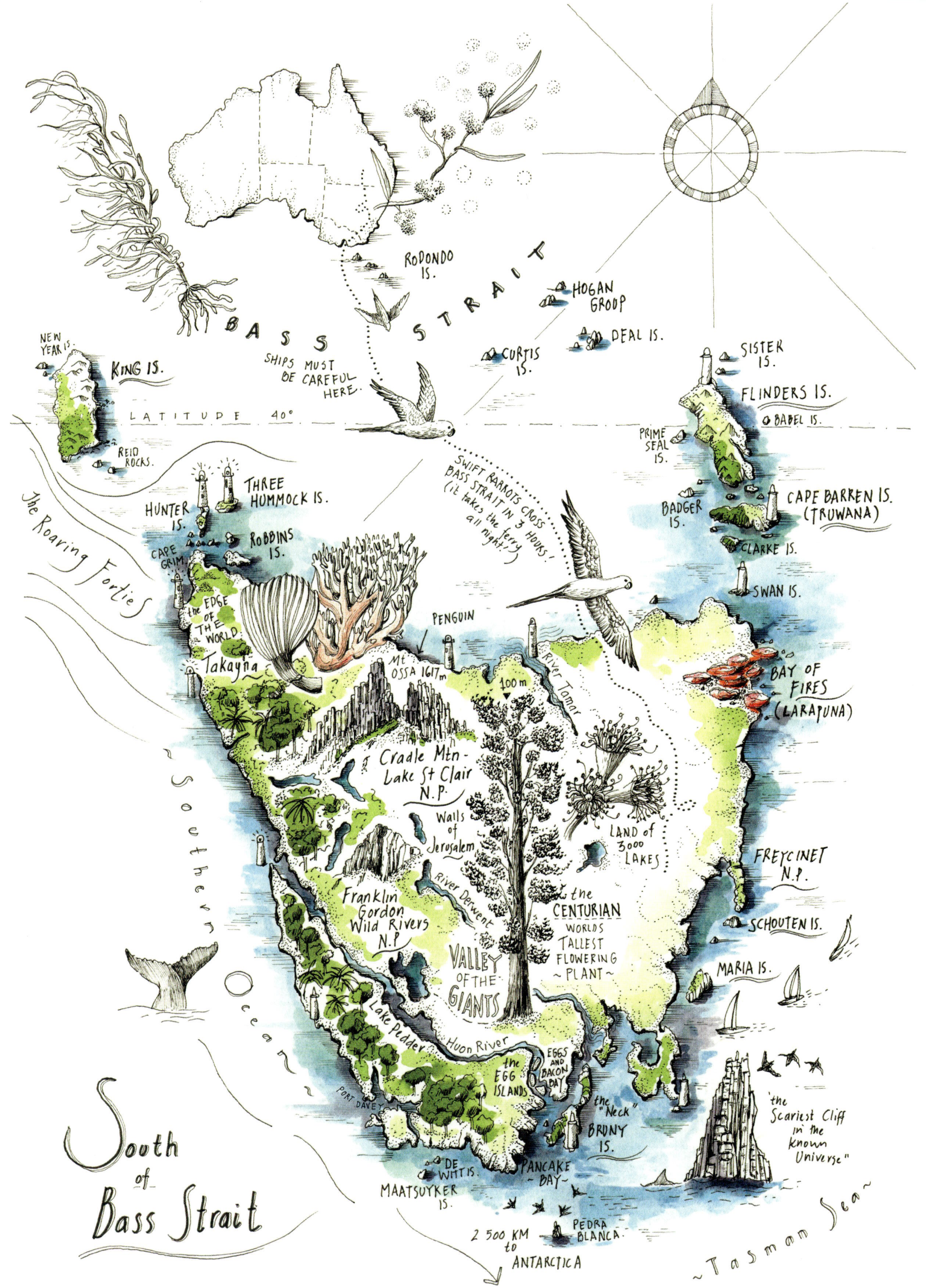

South of Bass Strait
BASS STRAIT
RODONDO IS.
HOGAN GROUP
DEAL IS.
CURTIS IS.
SHIPS MUST BE CAREFUL HERE.
NEW YEAR IS.
KING IS.
LATITUDE 40°
REID ROCKS.
SISTER IS.
FLINDERS IS.
BABEL IS.
PRIME SEAL IS.
BADGER IS.
CAPE BARREN IS. (TRUWANA)
CLARKE IS.
SWAN IS.
SWIFT PARROTS CROSS BASS STRAIT IN 3 HOURS! (it takes the ferry all night!)
THREE HUMMOCK IS.
HUNTER IS.
ROBBINS IS.
CAPE GRIM
The Roaring Forties
the EDGE OF THE WORLD
Takayna
PENGUIN
Mt OSSA 1617m
River Tamar
100m
BAY OF FIRES (LARAPUNA)
Cradle Mtn-Lake St Clair N.P.
Walls of Jerusalem
LAND of 3000 LAKES
FREYCINET N.P.
Franklin Gordon Wild Rivers N.P
River Derwent
the CENTURIAN WORLDS TALLEST FLOWERING ~PLANT~
SCHOUTEN IS.
VALLEY OF THE GIANTS
MARIA IS.
~Southern Ocean~
Lake Pedder
Huon River
EGGS AND BACON BAY
the EGG ISLANDS
PORT DAVEY
the "Neck"
BRUNY IS.
'the Scariest Cliff in the Known Universe"
DE WITT IS.
PANCAKE ~BAY~
MAATSUYKER IS.
PEDRA BLANCA
2 500 KM to ANTARCTICA
~Tasman Sea~

I sought to turn the tables with this map—showing the mainland unadorned and much smaller in scale than its southern island state. «

« South of Bass Strait

Alex Hotchin

To the south of mainland Australia, across the shallow Bass Strait, lies Tasmania. At less than one one-hundredth the size of its mainland neighbor, this island state is diminutive and has an unfortunate history of occasionally being left off maps of Australia. It is a curious omission, as Tasmania is considered one of the most beautiful parts of this southern land, abundantly covered with wild rivers and remnant old-growth forests. I sought to turn the tables with this map—showing the mainland unadorned and much smaller in scale than its southern island state. Less attention is taken in the drawing of the coastline of the mainland, in comparison to the highly detailed edges of Tasmania. It is an island full of many treasures and surprises, and I have sought to give the drawing a whimsical and adventure-island feel through its linework, color, and selective addition of unusual place-names. Here I describe just a few places and ways in which I sought to do so.

Tasmania is home to some of the tallest trees, fastest birds, and scariest cliffs in all the world, and the sizes of the drawings to represent these are overemphasized, to communicate their big personalities. Tasmania is brightly colored and teeming with notes and drawings of some of its impressive inhabitants and topographic features. The color palette is kept predominantly to green and blue—the former to emphasize the abundance of forest, and the latter, a blue of many shades, to bring attention to the treacherous seas. The sea color is kept tight to the edges of the island to emphasize the shape. A minimal use of additional color allows the oversize fungi and red rocks of the Bay of Fires to attract viewers' attention.

An underwater jungle of giant kelp inhabits Bass Strait. This shallow sea is therefore dangerous for the ships and ferries that cross. They must be careful. The passenger ferry takes all night to cross 350 km of water, but the fastest parrots on earth can do it in just three hours.

Nature *in the* CITY

What is San Francisco to you? The city is people, buildings, and views. It is also a peninsula made over millennia by rocks transported from far away and uplifted from deep below by powerful tectonic forces that continue to shape—and shake—our region today. Sunlight, weather, and geology create a biodiversity hotspot for more than 2000 species, some of which live here and only here. For countless generations, California Indians have tended the landscape in harmony with wild plants and animals. In the 1700s, European colonization radically altered many natural and cultural processes which today we are working to restore. The past helps us re-see the present, and move into a more harmonious future full of what Charles Darwin called endless forms, most beautiful: life.

EXPLORE AND DISCOVER

- NATURAL AREAS YOU CAN VISIT
- PARKS & OPEN SPACES
- TREE COVER
- LAKES & WATERWAYS TODAY
- HISTORIC WETLANDS & WATERWAYS
- BEACHES
- NATIVE PLANT NURSERIES
- NATURE EDUCATION SITES
- CAMPGROUNDS
- BART & CALTRAIN
- HISTORIC SHORELINE
- SELECT TRAILS

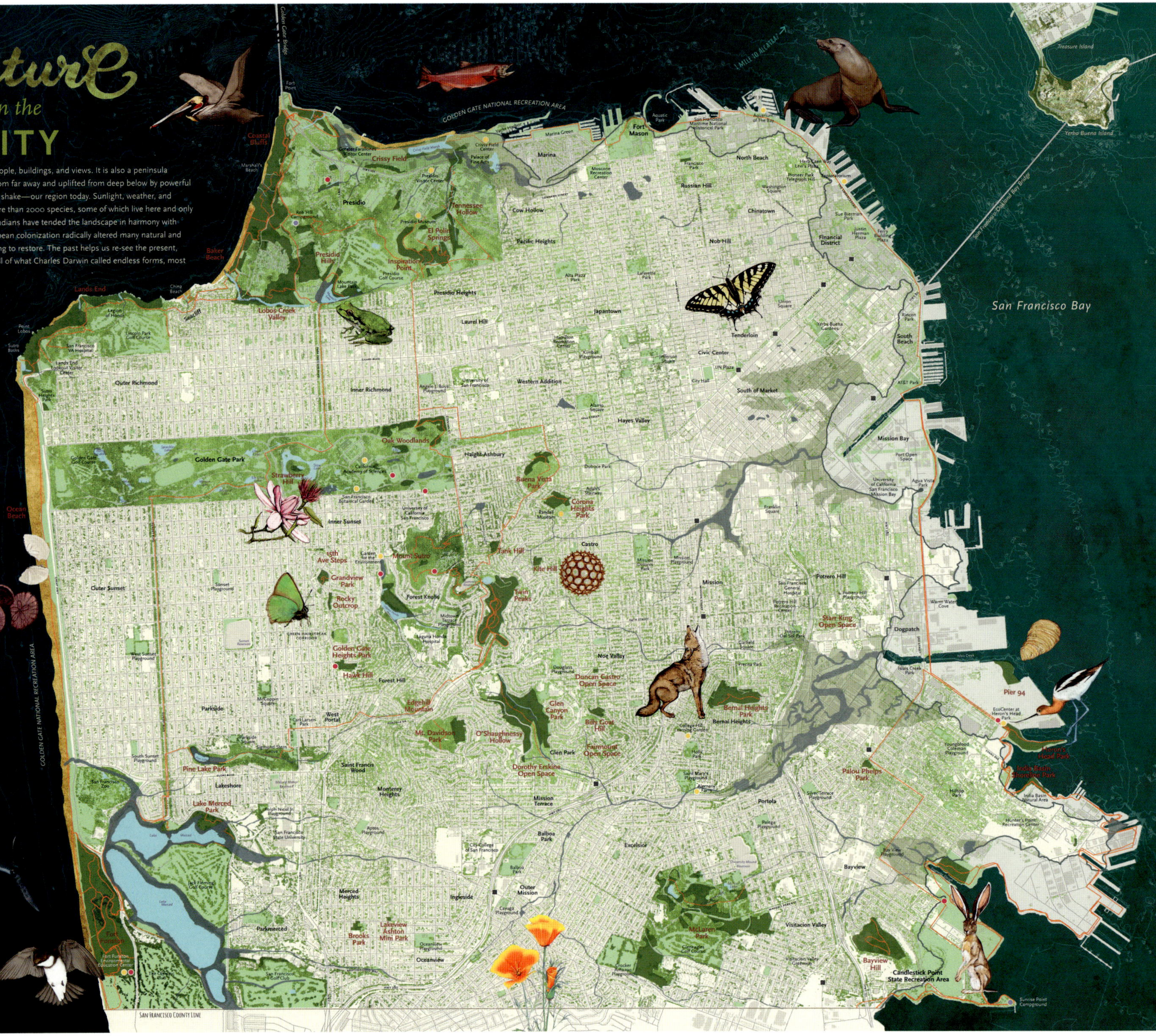

Have You Seen these Locals in the City

AMERICAN AVOCET
Recurvirostra americana
Winter-breeding avocets sport rusty red plumage on their necks and chests, which are gray and white in other seasons.
HERON'S HEAD PARK, YOSEMITE SLOUGH

BANK SWALLOW
Riparia riparia
In April and May, bank swallows dig nest burrows in tall, sandy cliffs. They perform vertical flying feat over the ocean.
FORT FUNSTON

BLACK-TAILED JACKRABBIT
Lepus californicus
Making a surprise return to the southeast corner of the city, the jackrabbit keeps cool by releasing he through its big ears.
CANDLESTICK POINT, HERON'S HEAD PARK

BLUE WHALE
Balaenoptera musculus
The largest animal on Earth feeds on some of the smallest, including zooplankton and krill, when the seasonal California Current brings nutrients from the deep ocean up to the surface.
OFFSHORE NEAR THE FARALLON ISLANDS

CALIFORNIA BROWN PELICAN
Pelecanus occidentalis californicus
Precise v-shaped squadrons of pelicans have cruise the shifting coastline for at least 30 million years.
COASTAL BLUFFS, OCEAN BEACH

CALIFORNIA POPPY
Eschscholzia californica
In 1903, botanist Sarah Plummer Lemmon designated the California poppy our state flower. The sunshine flower blooms in every corner of the city and across the state. Celebrate California Poppy Day on April 6.
CITYWIDE

CALIFORNIA SEA LION
Zalophus californianus
Who knows why? After the 1989 Loma Prieta earthquake a colony of sea lions began hauling up on Pier 39 and have barked there ever since.
PIER 39

CAMPBELL'S MAGNOLIA
Magnolia campbellii
The San Francisco Botanical Garden has the world most significant collection of magnolias outside Ch where most of the species originate.
GOLDEN GATE PARK AND CITYWIDE

CHINOOK SALMON
Oncorhynchus tshawytscha
Salmon swim upstream to spawn in woodland waters, where they die, thus providing nutrients that help the trees grow.
GOLDEN GATE THROUGH THE BAY–DELTA

COYOTE
Canis latrans
Among the most cosmopolitan of creatures, coyotes are equally at home in remote wilderness and in cities including San Francisco, where they have adapted to dense human presence.
BERNAL HEIGHTS PARK, GOLDEN GATE PARK, PRESI

GREEN HAIRSTREAK BUTTERFLY
Callophrys viridis
Since 2006, local school children and neighbors have worked with Nature in the City to create the Green Hairstreak Corridor, a series of connected habitats for this year-round resident of the city.
GRANDVIEW PARK, HAWK HILL, COASTAL BLUFFS

OLYMPIA OYSTER
Ostrea lurida
Oyster reefs provide tasty morsels and create a hard substrate that helps shorelines hold their own against sea level rise.
PIER 94, SAN FRANCISCO BAY

PACIFIC CHORUS FROG
Pseudacris regilla
Citywide restoration efforts have helped bring back this small native amphibian with a huge ribbiting voice.
PRESIDIO, MISSION DISTRICT, POTRERO HILL

RADIOLARIAN CHERT
Radiolaria
The skeletal remains of ancient zooplankton have undergone geological pressures through time and space to become reddish, ribbon-like rock formations today.
TWIN PEAKS, CORONA HEIGHTS, GLEN CANYON

SAND DOLLAR
Dendraster excentricus
Living sand dollars are covered with millions of tiny feet which help them stay put in pummelin waves. What you see on the beach is the animal's skeleton, or test.
OCEAN BEACH

WESTERN TIGER SWALLOWTAIL
Papilio rutulus
These hearty butterflies live along Market Street, where the urban landscape is somewhat like their native river canyon habitat.
FINANCIAL DISTRICT, MARKET STREET, GLEN PARK

Curious? Download the free iNaturalist app
Learn where species have been seen and share your own observation

Shaping the City

Earthquakes rumble and the continent shifts, bringing the rocks of yesterday to the trailheads of today. Ancient climatic cycles and tectonic grinding erode the rock into endemic soils. Plants that grow only in San Francisco sustain native species and make our city unique. High hills of bedrock push up through sand. Below-ground basins have captured rain for millennia, storing it, and feeding creeks.

Laurel Hill, Twin Peaks
Long-term Relationships
Radiolaria are the skeletons of tiny, ancient sea creatures that have sunk to the bottom of the ocean and been compressed into layered rock called chert. **Serpentinite** is a metamorphic rock to which plants like **Franciscan manzanita** are specially adapted. **Wrentits** have been eating manzanita berries for thousands of years. The root systems of native **purple needlegrass** extend deep into the soil, and some are hundreds of years old. **California poppies** give orange greetings to the day year after year after year.

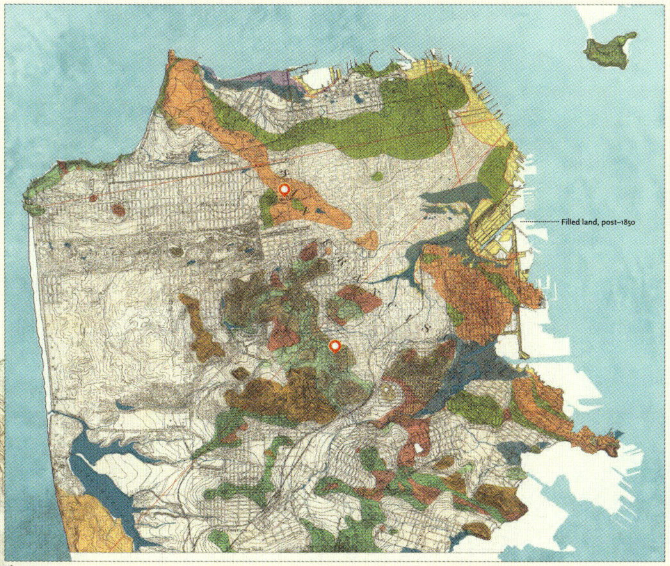

Shaping the City Map

In 1908 Andrew C. Lawson mapped ten geologic areas and the original shoreline of the peninsula for the Earthquake Investigation Commission. Waterways as they existed pre-urban development are illustrated in blue.

Soil Types & Waterways

- Made land
- Pleistocene sandstone (sand)
- Pliocene-Merced series sandstone
- Serpentinized peridotite (serpentinite)
- Spheroidal basalt

Franciscan Series

- Sandstone—stratigraphic
- Upper formation of radiolarian chert
- Middle formation of sandstone
- Lower formation of radiolarian chert
- Lower formation of sandstone
- Historic wetlands and waterways

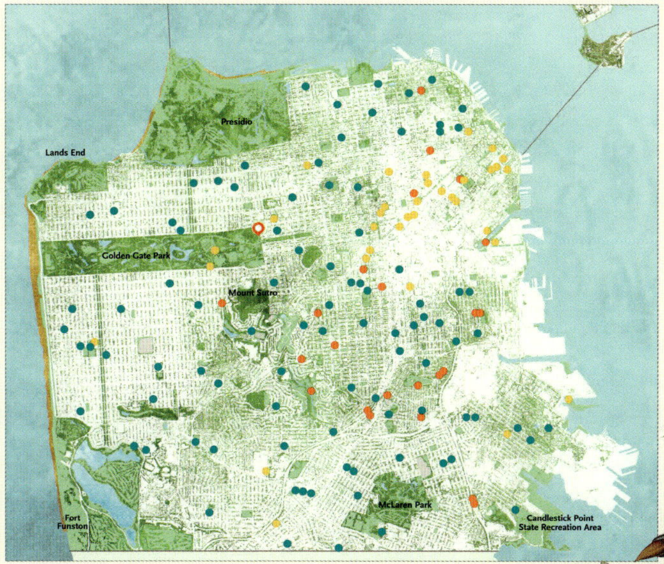

Gardening the City Map

Most city residents are within a 10-minute walk of the 230 parks and open spaces maintained by San Francisco Recreation and Parks Department. This map includes community gardens, green schoolyards, and street trees planted largely by Friends of the Urban Forest.

Tended Landscapes

- Community gardens
- Rooftop gardens
- Green schoolyards
- Parks and open spaces
- Tree cover

Gardening the City

San Francisco is naturally a coastal scrub, grassland, and dune ecosystem, and nature gardened our city first. For thousands of years, California Indians used fire, plantings, and other strategies to support wild species. European-Americans planted gardens like Golden Gate Park, replacing native habitat, but some native species have adapted to introduced plants.

Stanyan Street
Cycles of Life
The acorn-burying **Western Scrub-Jay** is responsible for planting virtually all our iconic **coast live oak** trees. The burrowing **eastern grey squirrel** turns over the soil, replants seeds, and provides living space for other species. Introduced **southern magnolia** thrive in our Mediterranean climate as does native **coast buckwheat**. Hundreds of people help restore host and nectar plants for the **green hairstreak butterfly**, while hearty **western tiger swallowtail butterflies** flitter alongside commuters, finding unlikely habitat in planters and corner parks downtown. Community and rooftop gardens provide filling stations for bugs, birds, and other species to help them on their way.

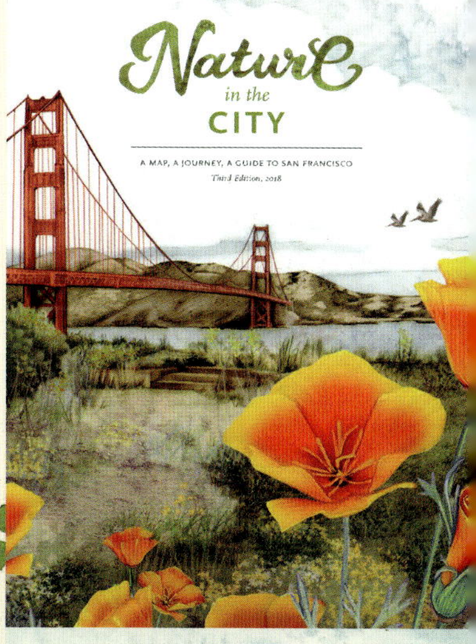

Many Stories, One City

This two-sided map offers five ways to see nature in San Francisco. The front presents a big picture of some species and habitats that help create a place of beauty and belonging. Four maps on the flip side explore some ways we are re-learning to co-exist here. From the Bayshore to Ocean Beach, darting through traffic, alighting on flowers, digging into the earth, and flying overhead, an amazing abundance of creatures share their life journeys with ours.

Resilient City

The city has developed, we have treated the shoreline as a boundary rather than the vital connection it is between water and land. We have moved it, filled it in, and paved over its sponge-like ability to buffer storm surge, sea level rise, and tidal events.

Heron's Head
Back to the Future
San Francisco is a leader in ecological restoration, including thousands of acres of tidal wetland. **Olympia oysters** and **pickleweed** help recreate healthy functioning that supports human systems coping with rapid change. We welcome back the **American Avocet,** and even the **black-tailed jackrabbit**. Once again, the Presidio hosts the **Pacific chorus frog**. "Trickster" to many Native Americans, the **coyote** followed humans to the Bay Area. Our top predator along with the occasional **mountain lion**, coyotes help keep rodent populations in check.

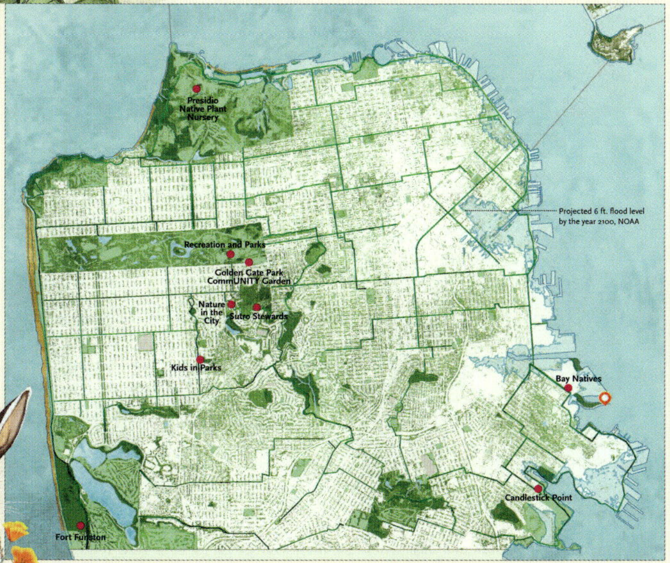

Resilient City Map

San Francisco's natural areas help create resilience to impacts from sea level rise and drought. They contain the city's remaining patches of native habitats and are managed for biological diversity. Tree cover connects parks and other habitats, as will 24 "Green Connections" routes when they are built along the city's streets. Sea level rise and flooding projections for the year 2100 are provided by the National Oceanic and Atmospheric Administration. Also shown are native plant nurseries (volunteers welcome). Use *sfplantfinder.org* to help support local biodiversity.

Landscapes and Changing Shorelines

- Natural areas
- Tree cover
- Native plant nurseries
- Green Connections Plan
- Future shoreline/sea level rise

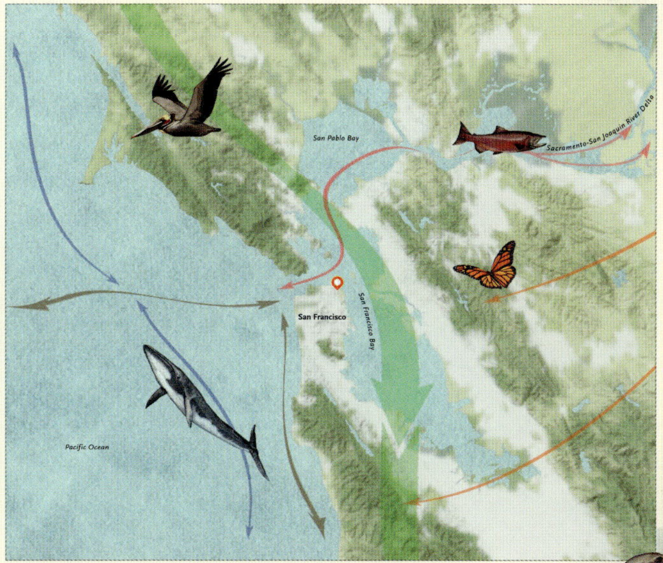

Connecting the City Map

Seasonal migrations include marine and terrestrial species that pass by, over, through, and around the Bay, connecting our city to regional, continental, and global systems.

Seasonal Migration Routes

- Blue whale
  Northern California *(summer–fall)*
  Mexico and South America *(winter–spring)*
- California sea lion
  Farallones and Channel Islands, breeding and birth *(spring–summer)*
  California coast *(fall–winter)*
- Chinook salmon
  Pacific Ocean to Sacramento-San Joaquin River Delta watershed
- Monarch butterfly (western U.S.)
  California coastal woodlands *(fall–winter)*
  California and western states *(spring–summer)*
- Pelican
  California coast and Channel Islands *(summer–fall)*
  Mexico *(winter–spring)*

Connecting the City

Natural connections play out over time, driven by seasonal, climatic, and geological forces that result in California's unique species. The Bay is part of a regional circulatory system, an estuary connecting the Pacific Ocean to the Delta made by the Sacramento and San Joaquin Rivers.

Pier 39
Here Comes the Sun
Summer ocean currents help create fog and fuel the marine food web, from tiny **krill** to migrating **blue whales.** In fall, young **Red-tailed Hawks** make their first journey south across the Golden Gateway. Some **California Brown Pelicans** hang out here all winter, while others migrate. **California sea lions** give birth on the southern coast, then come north. **Monarch butterflies** spend part of their four-generation migration here. **Chinook salmon** fulfill their life's journey by swimming under the Golden Gate Bridge through fresh, brackish, and saltwater, completing the cycle at the heart of the Bay-Delta.

Find *Your* Nature in the City Everyday

Be curious and explore
Share your photos on iNaturalist
Grow wildlife friendly plants
Volunteer at a local park

To learn more ways to engage, visit: *www.natureinthecity.org/map*

Acknowledgments

In the tradition of California Indians who have shaped San Francisco for thousands of years, we extend our gratitude to the Earth itself, and to the people who continue to teach us that stewardship is at the heart of our connection to nature.

Creative Team

Leah Elamin, *Graphic Designer*
Mary Ellen Hannibal, *Writer*
Exploratorium
Susan Schwartzenberg, *Content Advisor*
Ink Dwell Studio
Jane Kim, *Artist*
Lichen Projects
Lindsay Irving, *Project Manager & Cartography*
SF Planning Department
Michael Webster, *GIS & Cartography*

Map Team

California Academy of Sciences
Rebecca Johnson, Alison Young
Nature in the City
Amber Hasselbring, Bette McDonnell, Bill Brick, Ariel Wang
Presidio Trust
Damien Raffa
SF Department of the Environment
Peter Brastow, Luke Watcher
SF Public Library
Kelley Trahan
SF Recreation and Parks
Dylan Molnar

Credits

Andrew C. Lawson, 1908. Geological Map of San Francisco, David Rumsey Historical Map Collection.
All city and street related data, SF Planning Department.
Current open water, reservoirs and historic wetlands and waterways, SF Dept. of Public Works, SFPUC, Seep City Water History Map Joel Pomerantz. Published by Thinkwalks, 2015.
Natural Areas, Parks and Open Spaces, SF Dept. of the Environment and SF Recreation and Parks Dept.
Connecting the City terrain map, Stamen Design.

This map was made possible by a generous grant from **EJF Philanthropies**. Additional thanks go to the many individuals and organizations that provided guidance and resources to help make the map, including AMP Printing, Bay Nature Magazine, GreenInfo Network, Institute for Creative Integration, SmallWorks Framing, and Stamen Design.

silient *City*

city has developed, we have treated the shoreline as a
ary rather than the vital connection it is between water
d. We have moved it, filled it in, and paved over its
-like ability to buffer storm surge, sea level rise, and
l events.

s Head

the Future

ancisco is a leader in ecological restoration, including
nds of acres of tidal wetland. **Olympia oysters** and
eed help recreate healthy functioning that supports
systems coping with rapid change. We welcome back
erican Avocet, and even the **black-tailed jackrabbit**.
gain, the Presidio hosts the **Pacific chorus frog**.
ter" to many Native Americans, the **coyote** followed
s to the Bay Area. Our top predator along with the
onal **mountain lion**, coyotes help keep rodent
tions in check.

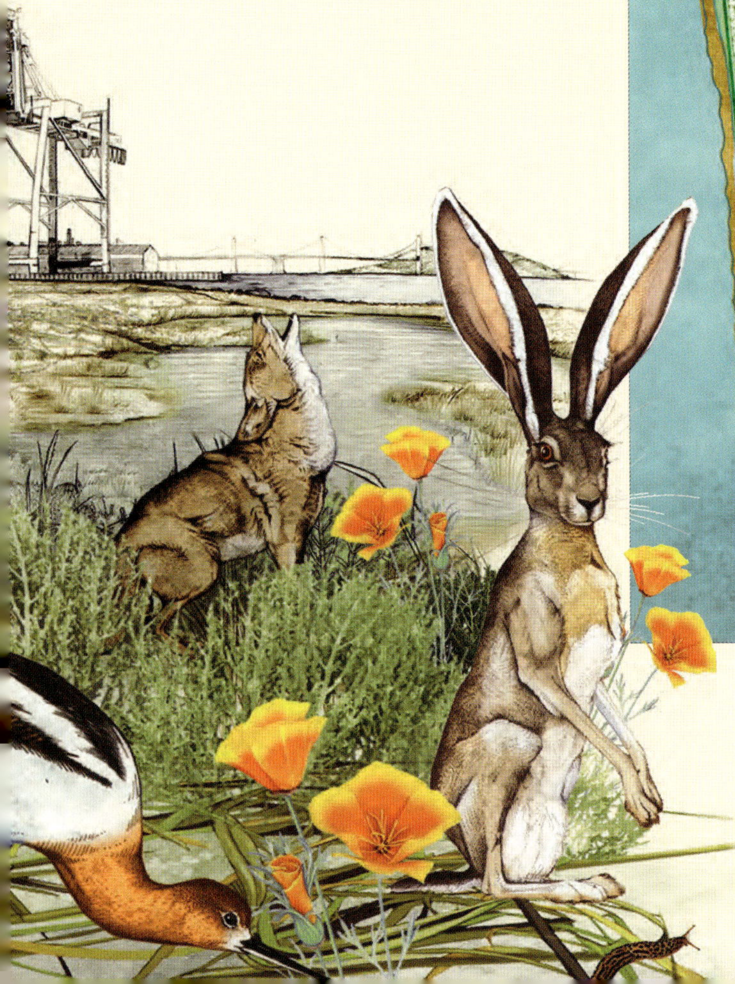

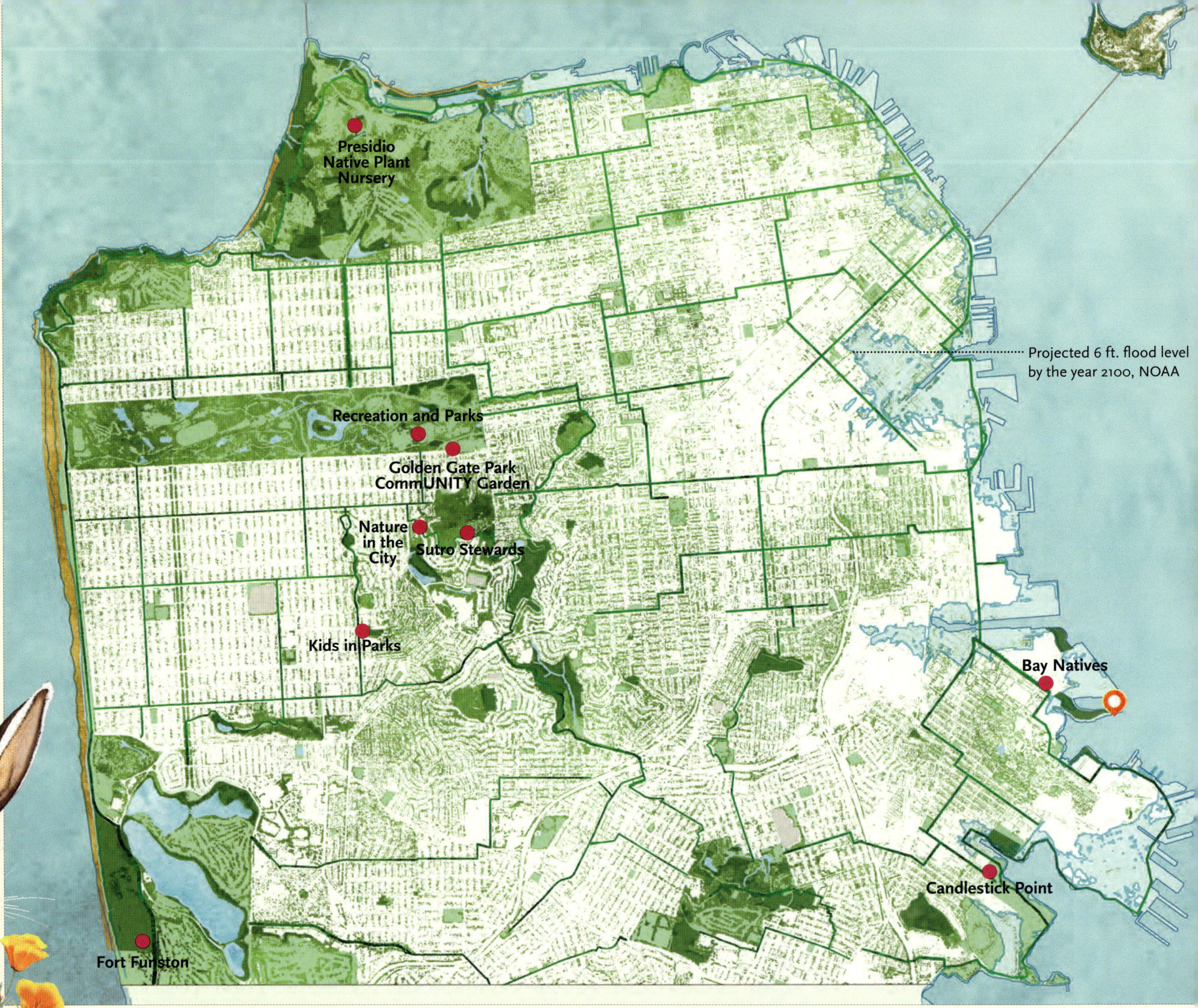

Resilient City Map

San Francisco's natural areas help create resilience to impacts from sea level rise and drought. They contain the city's remaining patches of native habitats and are managed for biological diversity. Tree cover connects parks and other habitats, as will 24 "Green Connections" routes when they are built along the city's streets. Sea level rise and flooding projections for the year 2100 are provided by the National Oceanic and Atmospheric Administration. Also shown are native plant nurseries (volunteers welcome). Use *sfplantfinder.org* to help support

Landscapes and Changing Shorelines

- Natural areas
- Tree cover
- Native plant nurseries
- Green Connections Plan
- Future shoreline/sea level rise

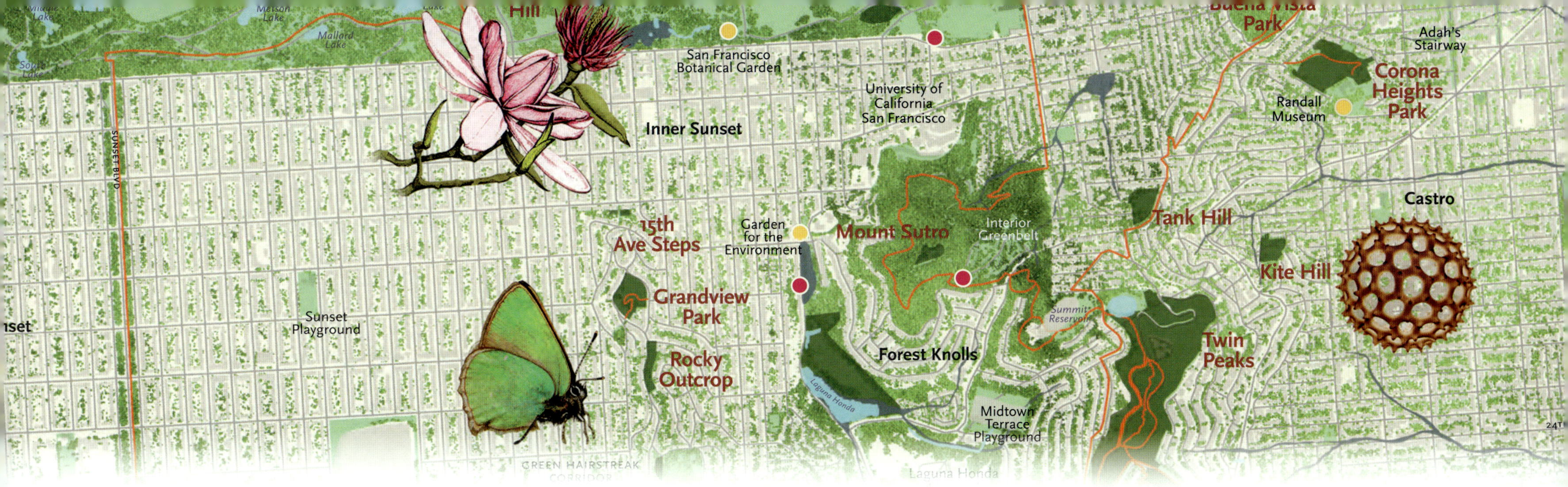

« Nature in the City: A Map, A Journey, A Guide to San Francisco

Lindsay Irving, Lichen Projects, cartography; Mary Ellen Hannibal, author; Amber Hasselbring, Nature in the City; Jane Kim, Ink Dwell Studio, artist

Earlier versions of the *Nature in the City* map go back nearly twenty years. The original maps showed open spaces in San Francisco, but the maps were mostly gray. They seemed to consider concrete first, and biodiversity second.

Our stalwart core of interdisciplinary mapmakers decided to radically revision the map. We invited folks from multiple city and cultural institutions, as well as independent artists, to reboot the very concept of "nature in the city." Inspired by early pictorial maps, we transformed more than fifty raw datasets generated by public agencies into hand-illustrated map layers. Leaning into art, we wanted to inspire a new way of looking at the role of humans in the history of the city and suggest our active role in shaping its future: a "creative emergent" vision of our city.

The work was exciting and prolonged (three years!). Our goal was to include past, present, and future narratives illustrating how those gray streets and buildings came to exist within San Francisco's 7×7-mile peninsula. We wanted to show that nature persists right in our backyards as well as within city parklands. We transformed tree canopy data taken from satellite imagery into a textured layer of green. This became our new base map.

Some truths here are ugly. Spanish missionaries, Gold Rush seekers, railroad barons, and others disrupted California Indian lifeways that originally sustained the ecosystem: coastal dunes hundreds of feet high were converted to a colonial idea of parks, causing local species extinctions; meandering waterways and fluid shorelines were drained and made rigid. We showed those layers, too. Yet we wanted to celebrate San Francisco's leadership in urban restoration and to appreciate native plants, pollinators, birds, and other original inhabitants of this place—who still live here!

The front of the map depicts a big picture of nature in the city. The back portrays four stories that highlight process and time: a shaped city engineered by tectonic, ocean, and human forces; a gardened city tended by ideals old and new; a creative city where landscapes and species are restored; and a connected city through which historic wildlife and water corridors intersect with contemporary movement pathways.

It has never been more urgent to see, feel, and interact with nature. Our map is a guide to past, present, and future. It is a map of the city in nature.

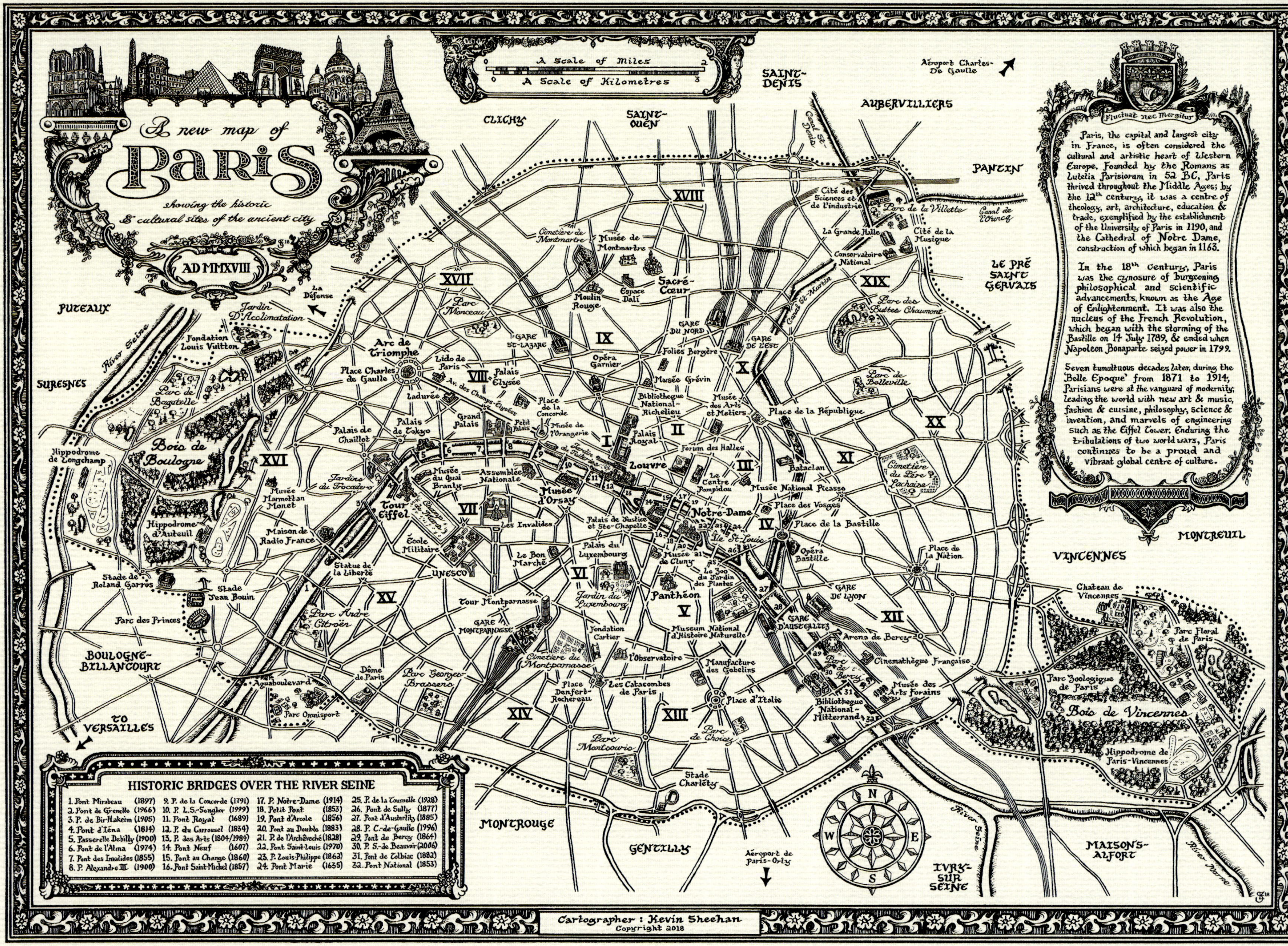
A new map of
Paris
showing the historic & cultural sites of the ancient city
AD MMXVIII
A Scale of Miles
A Scale of Kilometres
Aéroport Charles-De Gaulle
Fluctuat nec Mergitur
Paris, the capital and largest city in France, is often considered the cultural and artistic heart of Western Europe. Founded by the Romans as Lutetia Parisiorum in 52 BC, Paris thrived throughout the Middle Ages; by the 12th century, it was a centre of theology, art, architecture, education & trade, exemplified by the establishment of the University of Paris in 1190, and the Cathedral of Notre Dame, construction of which began in 1163.
In the 18th century, Paris was the cynosure of burgeoning philosophical and scientific advancements, known as the Age of Enlightenment. It was also the nucleus of the French Revolution, which began with the storming of the Bastille on 14 July 1789, & ended when Napoleon Bonaparte seized power in 1799.
Seven tumultuous decades later, during the 'Belle Epoque' from 1871 to 1914, Parisians were at the vanguard of modernity, leading the world with new art & music, fashion & cuisine, philosophy, science & invention, and marvels of engineering such as the Eiffel Tower. Enduring the tribulations of two world wars, Paris continues to be a proud and vibrant global centre of culture.
SAINT-DENIS
AUBERVILLIERS
CLICHY
SAINT-OUEN
PANTIN
LE PRÉ SAINT GERVAIS
PUTEAUX
SURESNES
BOULOGNE-BILLANCOURT
TO VERSAILLES
MONTROUGE
GENTILLY
IVRY-SUR-SEINE
MAISONS-ALFORT
VINCENNES
MONTREUIL
River Seine
La Défense
Jardin D'Acclimatation
Fondation Louis Vuitton
Parc de Bagatelle
Bois de Boulogne
Hippodrome de Longchamp
Hippodrome d'Auteuil
Stade de Roland Garros
Stade Jean Bouin
Parc des Princes
Arc de Triomphe
Place Charles de Gaulle
Lido de Paris
Av. des Champs-Elysées
Palais Elysée
Ladurée
Grand Palais
Petit Palais
Place de la Concorde
Musée de l'Orangerie
Palais de Tokyo
Palais de Chaillot
Jardins du Trocadéro
Musée Marmottan Monet
Maison de Radio France
Tour Eiffel
Champ de Mars
Musée du Quai Branly
Assemblée Nationale
Musée d'Orsay
Les Invalides
École Militaire
UNESCO
Statue de la Liberté
Parc André Citroën
Tour Montparnasse
Gare Montparnasse
Le Bon Marché
Palais du Luxembourg
Jardin du Luxembourg
Fondation Cartier
Cimetière du Montparnasse
Dôme de Paris
Parc Georges-Brassens
Aquaboulevard
Parc Omnisport
Place Denfert-Rochereau
Les Catacombes de Paris
l'Observatoire
Parc Montsouris
Stade Charléty
Place d'Italie
Parc de Choisy
Manufacture des Gobelins
Museum National d'Histoire Naturelle
Panthéon
Musée de Cluny
Le Jardin des Plantes
Palais de Justice et Ste-Chapelle
Notre-Dame
Île St-Louis
Louvre
Jardin des Tuileries
Palais Royal
Forum des Halles
Le Centre Pompidou
Musée National Picasso
Place des Vosges
Place de la Bastille
Opéra Bastille
Gare d'Austerlitz
Gare de Lyon
Arena de Bercy
Parc de Bercy
Cinémathèque Française
Bibliothèque Nationale-Mitterrand
Musée des Arts Forains
Place de la Nation
Cimetière du Père Lachaise
Bataclan
Place de la République
Musée des Arts et Métiers
Musée Grévin
Bibliothèque National-Richelieu
Opéra Garnier
Folies Bergère
Gare St-Lazare
Parc Monceau
Moulin Rouge
Espace Dalí
Sacré-Cœur
Musée de Montmartre
Cimetière de Montmartre
Gare du Nord
Gare de l'Est
Canal St-Martin
Parc de Belleville
Parc des Buttes Chaumont
Conservatoire National
La Grande Halle
Cité des Sciences et de l'Industrie
Parc de la Villette
Cité de la Musique
Canal de l'Ourcq
Canal St Denis
Chateau de Vincennes
Parc Floral de Paris
Parc Zoologique de Paris
Bois de Vincennes
Hippodrome de Paris-Vincennes
Aéroport de Paris-Orly
I II III IV V VI VII VIII IX X XI XII XIII XIV XV XVI XVII XVIII XIX XX
N E S W
HISTORIC BRIDGES OVER THE RIVER SEINE
1. Pont Mirabeau (1897)
2. Pont de Grenelle (1966)
3. P. de Bir-Hakeim (1905)
4. Pont d'Iéna (1814)
5. Passerelle Debilly (1900)
6. Pont de l'Alma (1974)
7. Pont des Invalides (1855)
8. P. Alexandre III (1900)
9. P. de la Concorde (1791)
10. P. L.-S.-Senghor (1999)
11. Pont Royal (1689)
12. P. du Carrousel (1834)
13. P. des Arts (1804/1984)
14. Pont Neuf (1607)
15. Pont au Change (1860)
16. Pont Saint-Michel (1857)
17. P. Notre-Dame (1914)
18. Petit Pont (1853)
19. Pont d'Arcole (1856)
20. Pont au Double (1883)
21. P. de l'Archêveché (1828)
22. Pont Saint-Louis (1970)
23. P. Louis-Philippe (1862)
24. Pont Marie (1635)
25. P. de la Tournelle (1928)
26. Pont de Sully (1877)
27. Pont d'Austerlitz (1885)
28. P. C.-de-Gaulle (1996)
29. Pont de Bercy (1864)
30. P. S.-de Beauvoir (2006)
31. Pont de Tolbiac (1882)
32. Pont National (1853)
Cartographer : Kevin Sheehan
Copyright 2018

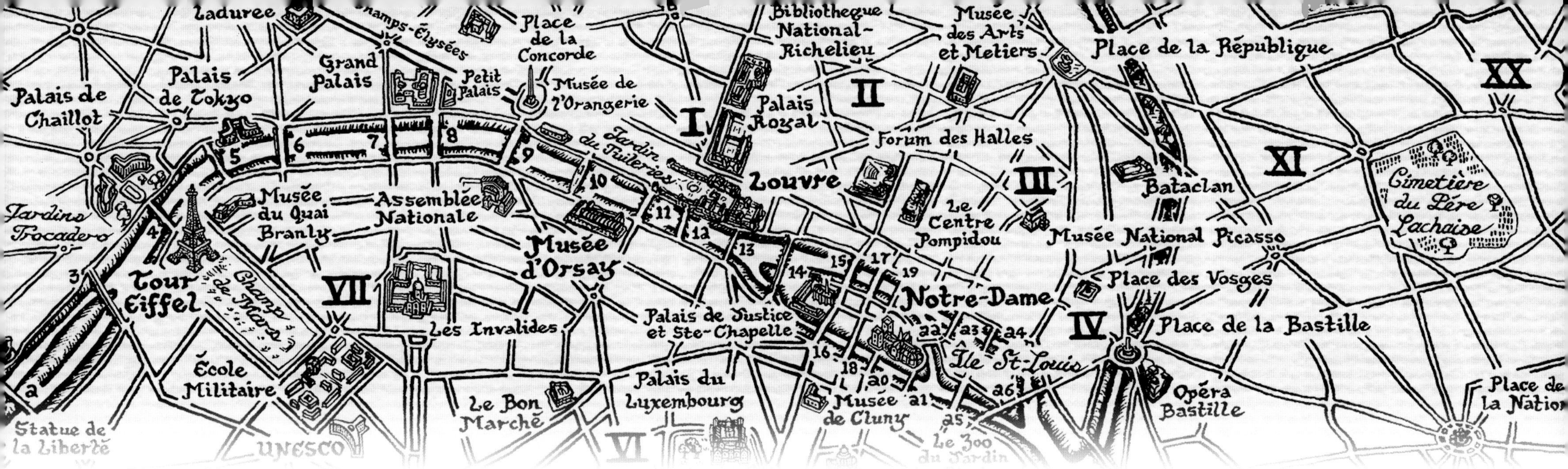

《 The Paris Map

Kevin Sheehan — Manuscript Maps

Paris: "La Ville Lumière." The city has been mapped innumerable times throughout history. I drew my Paris map in 2018, and it had to be exceptional: while I would sell prints through my business, Manuscript Maps, I planned to give the original to my sister as a combined twenty-first birthday and Christmas gift. She visited Paris with me for the first time in 2014 and absolutely loved it. I wanted to give her a map that could become a valued heirloom for generations.

Requiring a special aesthetic design to make the map appear like a centuries-old antique, I pored over hundreds of pictures of French maps for inspiration. Floral foliage motifs were commonplace, so I followed suit. I added ornamental cartouches for the title, for a brief history, and for the bridges because their toponyms could not fit onto the map itself. I incorporated seven of the most famous Parisian landmarks into the title cartouche because the oblique buildings on the map itself were too small to show much detail. As cartographers did centuries ago, I drew the map using dip pens with calligraphic and drawing nibs, and black India ink, onto calf vellum. The distinctive texture and waviness of vellum—which changed from day to day depending on humidity and temperature—posed interesting challenges.

All cartography blends science and art, in differing proportions. Most maps nowadays are created digitally: vectors, strings of code, hexadecimal colors, and font packages combine to create precise, uniform cartography. But to appear coherent and balanced, they still require the art of good design. While I strive for geographic accuracy, my maps are more aesthetic than scientific, and to that end I practice and advocate the heritage of drawing maps by hand. There is a depth of beauty and interest gained when every letter's form is slightly different from every other because of how the pen was held and how the ink flowed from the nib, and when every one of thousands of lines is not evenly spaced or evenly thick due to variations in pen pressure, humidity, flow, and vibrations in the hand. It is the accumulation of these imperfections, I believe, that turns maps into beautiful art.

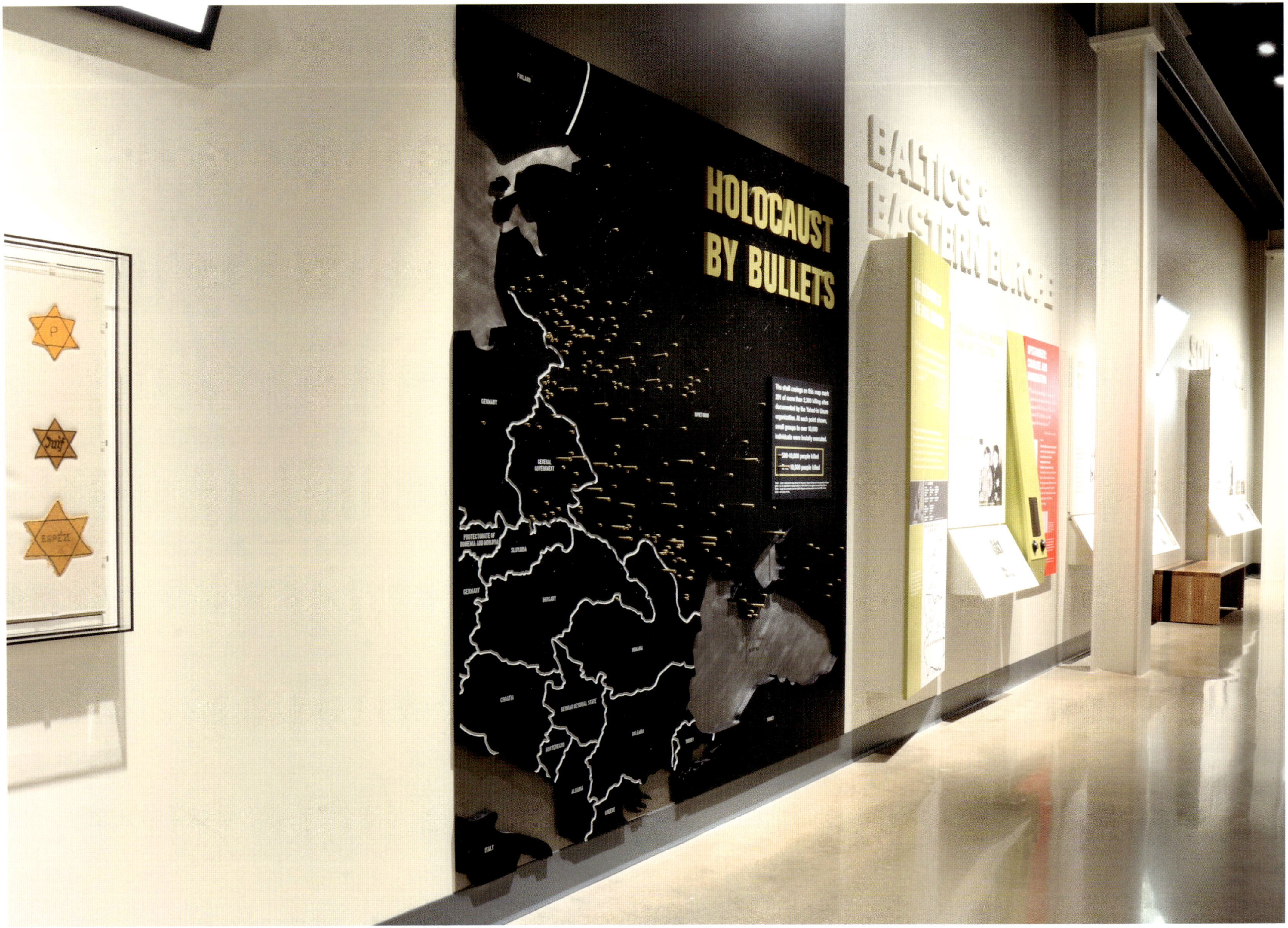
HOLOCAUST
BY BULLETS
BALTICS &
EASTERN EUROPE

《 Holocaust by Bullets

Edward Jacobs and Michael Berenbaum, Berenbaum Jacobs Associates, concept and design; Alison DeGraff Ollivierre, Tombolo Maps & Design, cartographer; Pacific Studio, fabrication

As part of an exhibit at the Dallas Holocaust and Human Rights Museum, we sought to create a map that would become a visual mnemonic for the heinous but little-known chapter of the Holocaust referred to as the "Holocaust by Bullets."

The "Holocaust by Bullets" was carried out by the *Einsatzgruppen* (Mobile Killing Units)—squads of German *Schutzstaffel* (SS), police personnel, and enthusiastic local militias that followed the German army through the invasion of the Soviet Union in June 1941. Their purpose was to "murder Jews and deprive them of their property." Victims also included Roma (Gypsies), officials of the Soviet Communist Party, and thousands of residents of institutions for people with mental and physical disabilities. The *Einsatzgruppen* shot any Jewish men, women, and children they encountered, and buried them in mass graves. Although an exact figure will never be known, approximately 1.5 million people were murdered by the *Einsatzgruppen*.

The organization *Yahad-In Unum* (whose mission is to identify sites of mass execution through extensive research and forensic evidence and spread awareness of the need to recognize and denounce the ongoing epidemic of global genocide) provided precise GPS data for the map.

In a museum, the challenge for the designer is to find modalities of presentation that capture visitors' curiosity and support the message of the map itself. The map is relatively free of text, offering only national borders for context. It was important to us to convey the tremendous geographic breadth of these operations. Against the harsh texture of black-painted particleboard, the shell casings create a visceral representation. We chose real shell casings to mark 10 percent of the 2,300 forensically identified sites of murder. Smaller shell casings mark the sites where fewer than 10,000 people were killed, and larger casings mark those where more than 10,000 were murdered. The effect is magnified by a literal shower of bullets hanging just above the map from the ceiling of the gallery (not visible in these photos).

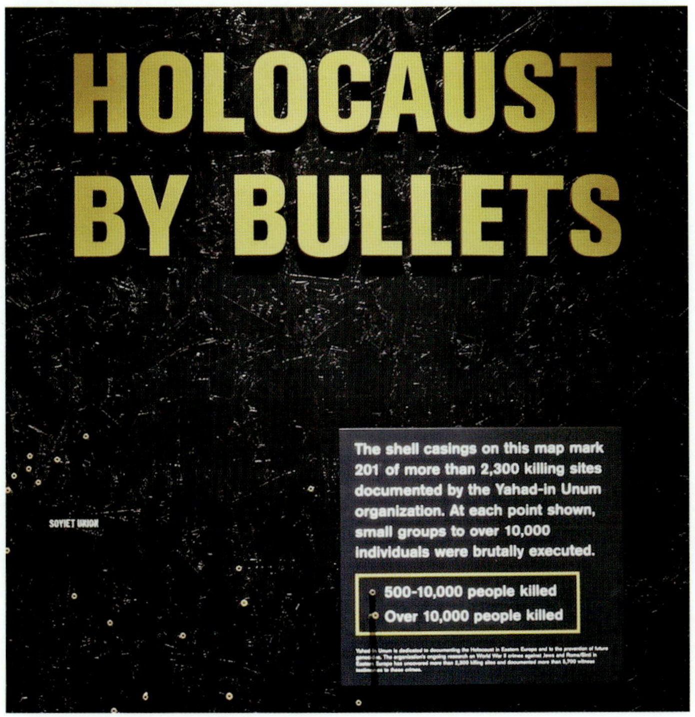

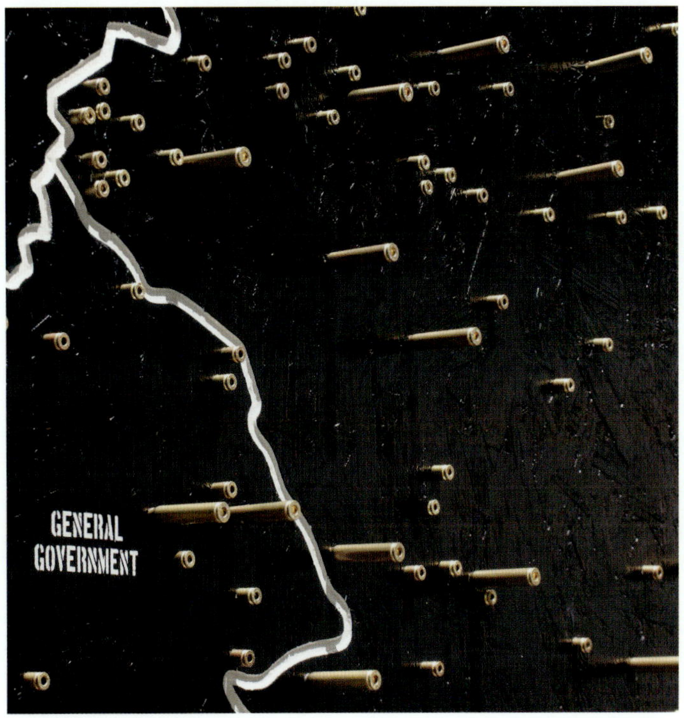

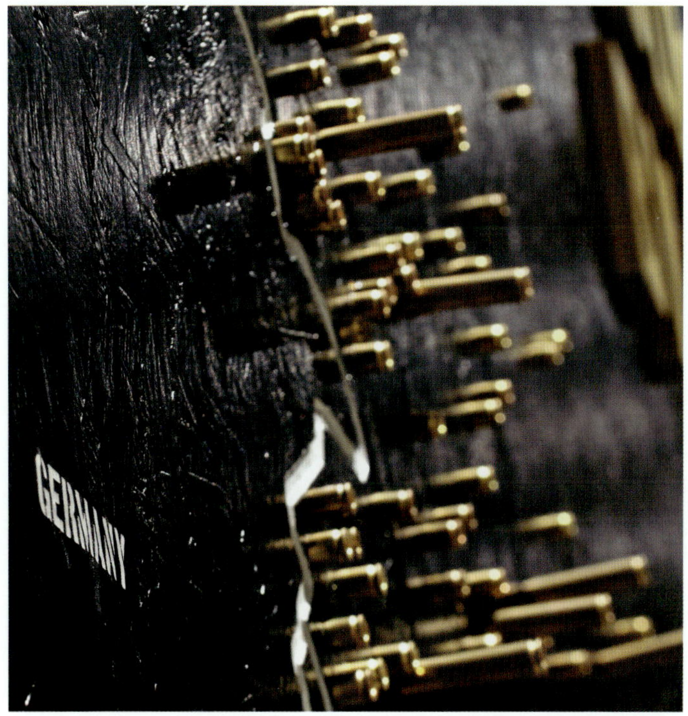

The Stikine Icefield »

Eric Knight

Panoramic maps showcase outrageous, borderline-impossible vistas in an idealized, photorealistic style that the viewer can instantly understand. The oblique perspective gives cartographers opportunities to use some of the same visual cues that humans rely on in our daily lives, so that orienting ourselves in this unknown place can feel easy and familiar.

I've gradually developed my own take on this style over the years. The fundamental techniques have changed relatively little since I first encountered Tom Patterson's descriptions of a workflow built around early 3D rendering software and Photoshop for both pre- and post-processing of raster data.

While this approach does rely on good-quality renderings, it's not until they are brought into Photoshop that any real control over design and style emerges. Incorporating a few newer tools into the workflow has helped automate some of the most time-consuming manual tasks, but has also created additional steps and transitions between formats that must be planned for and managed. The style you see here is not so much the result of a perfected sequence of steps, as the result of having a clear design goal, then getting to know all of the individual visual ingredients that might contribute to it, and coming up with ways to control them individually.

The Stikine Icefield is an unusually moody and evocative panorama, reflecting southeast Alaska's overcast, maritime climate. As always, my approach relies on using smooth gradients of color and value to create the illusion of depth, while punctuating those gradients with localized areas of higher contrast to accentuate particular features. Labels are subjected to the same kinds of transitions. Although this map is not shy when it comes to color, the gradations between those colors are subtle. The overall color palette remains cohesive, in part due to the muting and equalizing effects of distance.

This map also satisfies another important criterion: it gives a sense of the scale and complexity of the landscape without being especially large or complex itself. Most of my design choices have to do with making the landscape more legible, so that it reads at a glance, but also rewards deeper exploration. No matter the tools, the goal remains the same: to make immersive, hyperrealistic maps that blur the line between cartography and popular landscape art.

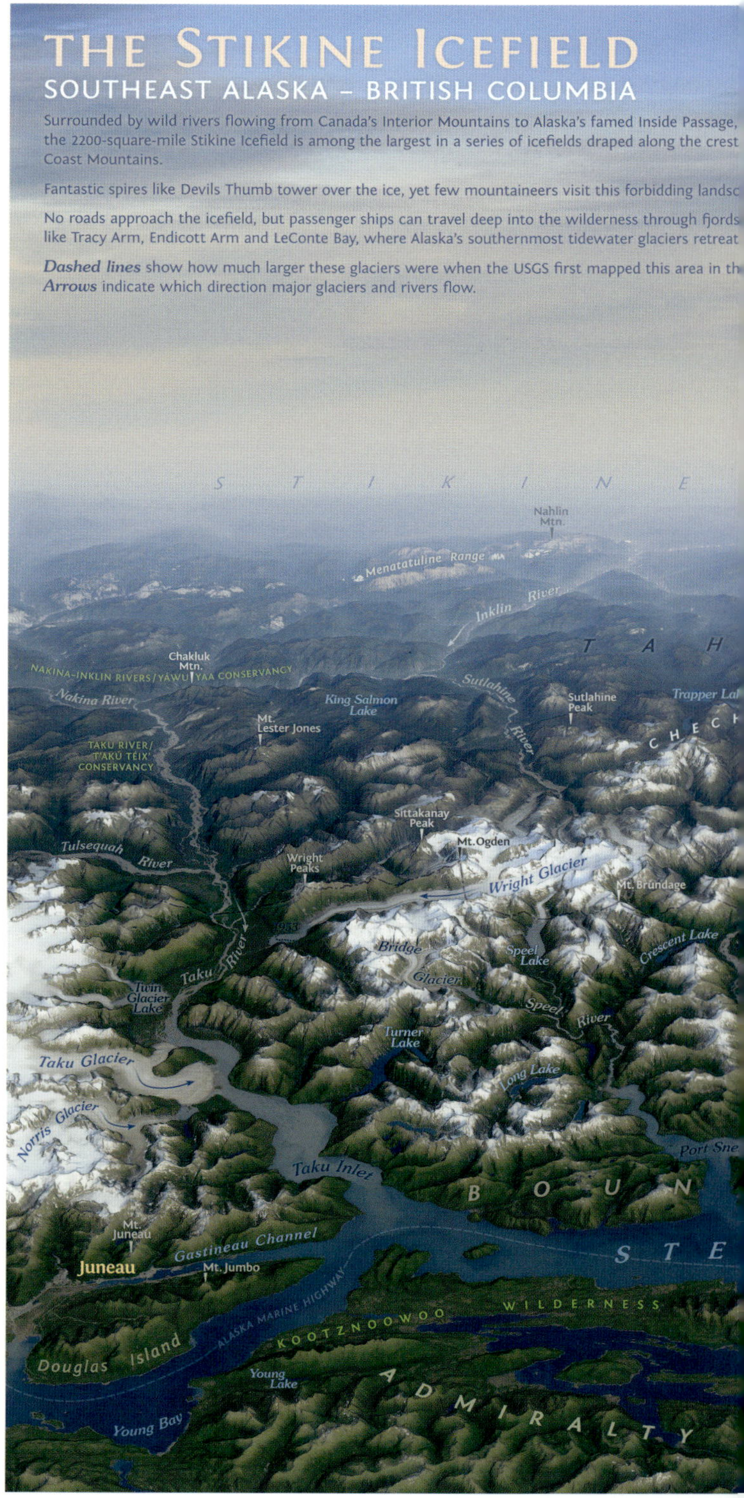

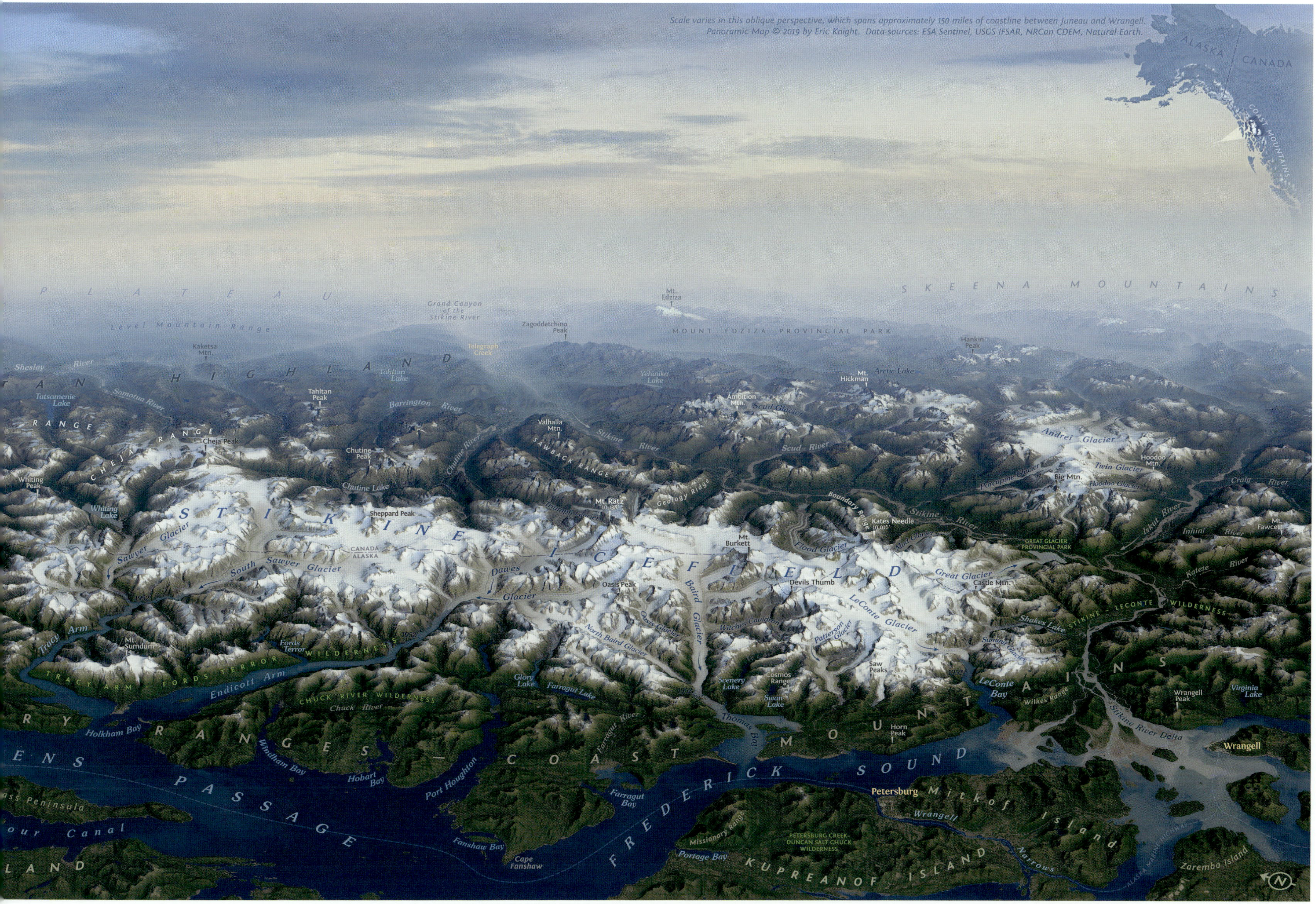

Scale varies in this oblique perspective, which spans approximately 150 miles of coastline between Juneau and Wrangell.
Panoramic Map © 2019 by Eric Knight. Data sources: ESA Sentinel, USGS IFSAR, NRCan CDEM, Natural Earth.
ALASKA
CANADA
COAST MOUNTAINS
PLATEAU
SKEENA MOUNTAINS
Mt. Edziza
Grand Canyon of the Stikine River
Level Mountain Range
Zagoddetchino Peak
MOUNT EDZIZA PROVINCIAL PARK
Hankin Peak
Kaketsa Mtn.
HIGHLAND
Telegraph Creek
Tahltan Lake
Sheslay River
Tatsamenie Lake
Samotua River
Tahltan Peak
Barrington River
Yehiniko Lake
Mt. Hickman
Arctic Lake
Ambition Mtn.
Scud Glacier
RANGE
CHEJA RANGE
Cheja Peak
Chutine Peak
Valhalla Mtn.
SAWBACK RANGE
Stikine River
Scud River
Andrei Glacier
Hoodoo Mtn.
Twin Glacier
Porcupine Glacier
Big Mtn.
Hoodoo Glacier
Craig River
Whiting Peak
Chutine Lake
Chutine River
Mt. Ratz
Geology Ridge
Boundary Range
Whiting Lake
Sheppard Peak
STIKINE ICEFIELD
Kates Needle
Iskut River
Inhini River
Mt. Fawcett
Sawyer Glacier
CANADA
ALASKA
Mt. Burkett
Flood Glacier
GREAT GLACIER PROVINCIAL PARK
South Sawyer Glacier
Dawes Glacier
Oasis Peak
Baird Glacier
Devils Thumb
Great Glacier
Castle Mtn.
Katete River
Tracy Arm
LeConte Glacier
Mt. Sumdum
North Baird Glacier
Witches Cauldron
Patterson Glacier
Shakes Lake
STIKINE - LECONTE WILDERNESS
Fords Terror
WILDERNESS
TRACY ARM - FORDS TERROR
Endicott Arm
Glory Lake
Farragut Lake
Scenery Lake
Cosmos Range
Saw Peaks
LeConte Bay
Wilkes Range
Wrangell Peak
Virginia Lake
CHUCK RIVER WILDERNESS
Chuck River
Swan Lake
Holkham Bay
Thomas Bay
Horn Peak
Stikine River Delta
Wrangell
RANGES - COAST MOUNTAINS
Windham Bay
Hobart Bay
Farragut River
Port Houghton
Farragut Bay
FREDERICK SOUND
Petersburg
Mitkof Island
STEPHENS PASSAGE
Peninsula
Canal
Fanshaw Bay
Cape Fanshaw
Missionary Range
PETERSBURG CREEK-DUNCAN SALT CHUCK WILDERNESS
Wrangell Narrows
Portage Bay
KUPREANOF ISLAND
ALASKA MARINE HIGHWAY
Zarembo Island
LAND
N

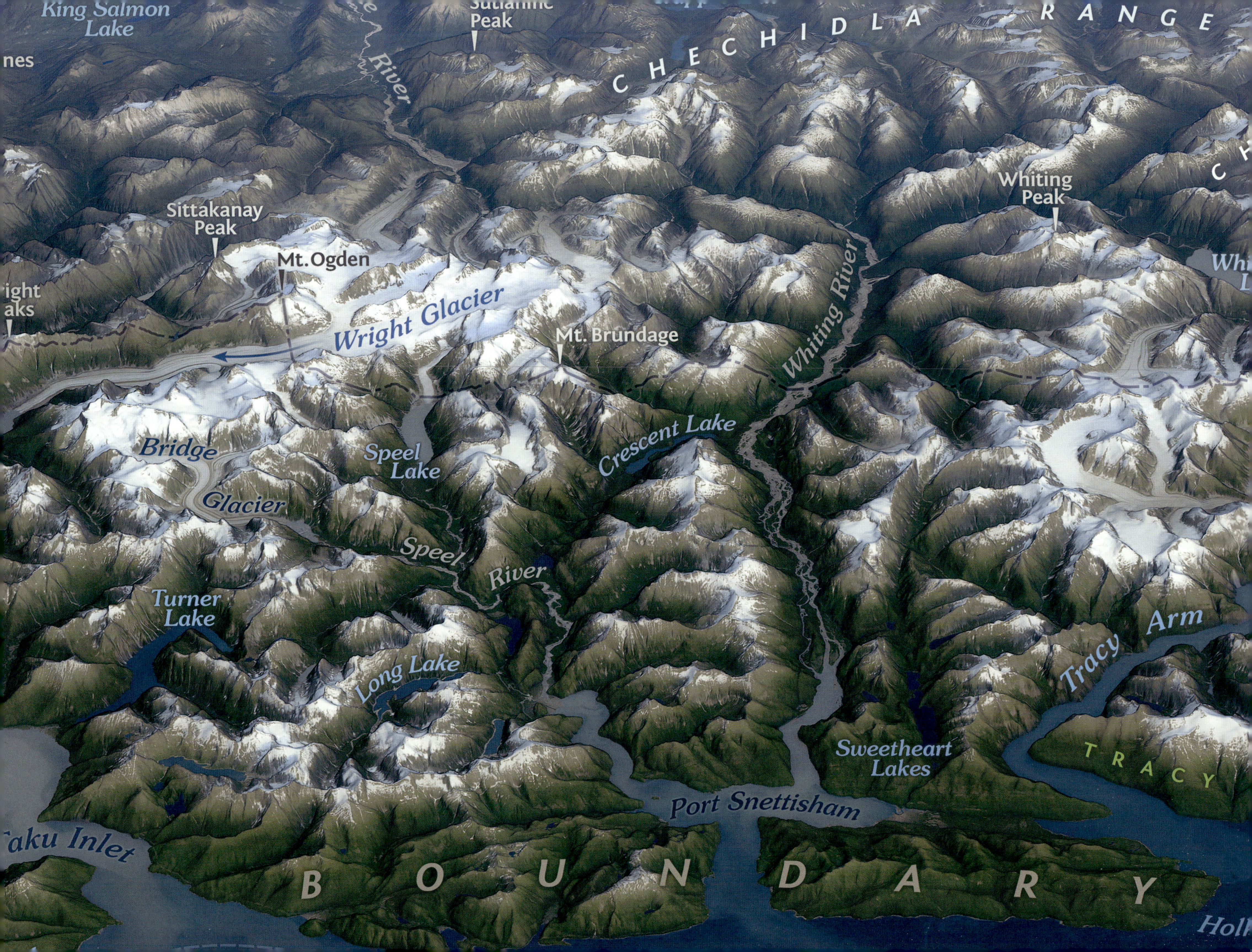
King Salmon Lake
River
CHECHIDLA RANGE
Whiting Peak
Sittakanay Peak
Mt. Ogden
Wright Glacier
Mt. Brundage
Whiting River
Bridge Glacier
Speel Lake
Crescent Lake
Speel River
Turner Lake
Long Lake
Tracy Arm
Sweetheart Lakes
TRACY
Port Snettisham
BOUNDARY

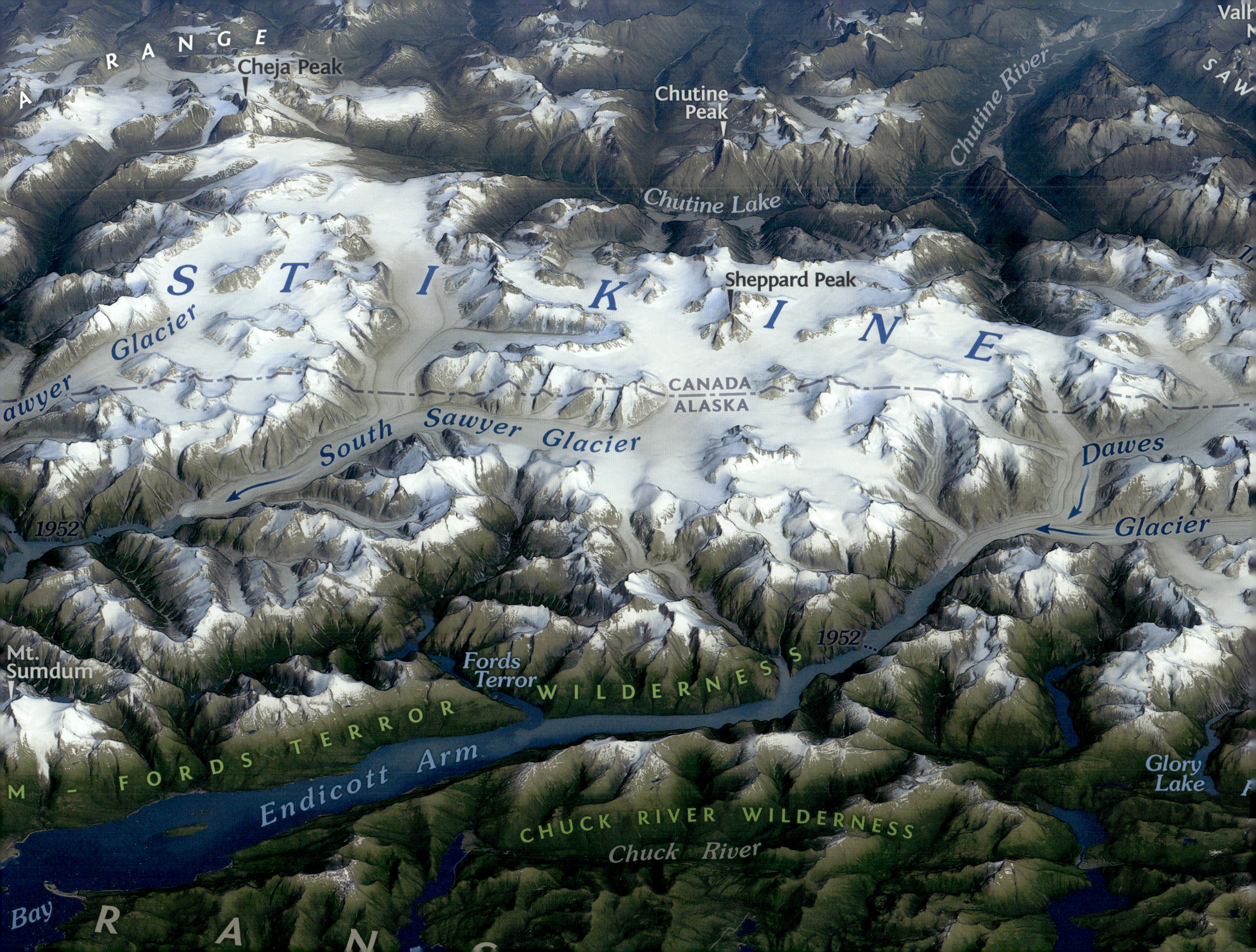

R A N G E
Cheja Peak
Chutine Peak
Chutine River
Chutine Lake
S T I K I N E
Sheppard Peak
Glacier
awyer
CANADA
ALASKA
South Sawyer Glacier
Dawes
Glacier
1952
1952
Mt. Sumdum
Fords Terror
WILDERNESS
M - FORDS TERROR
Endicott Arm
CHUCK RIVER WILDERNESS
Chuck River
Glory Lake
Bay
R A N

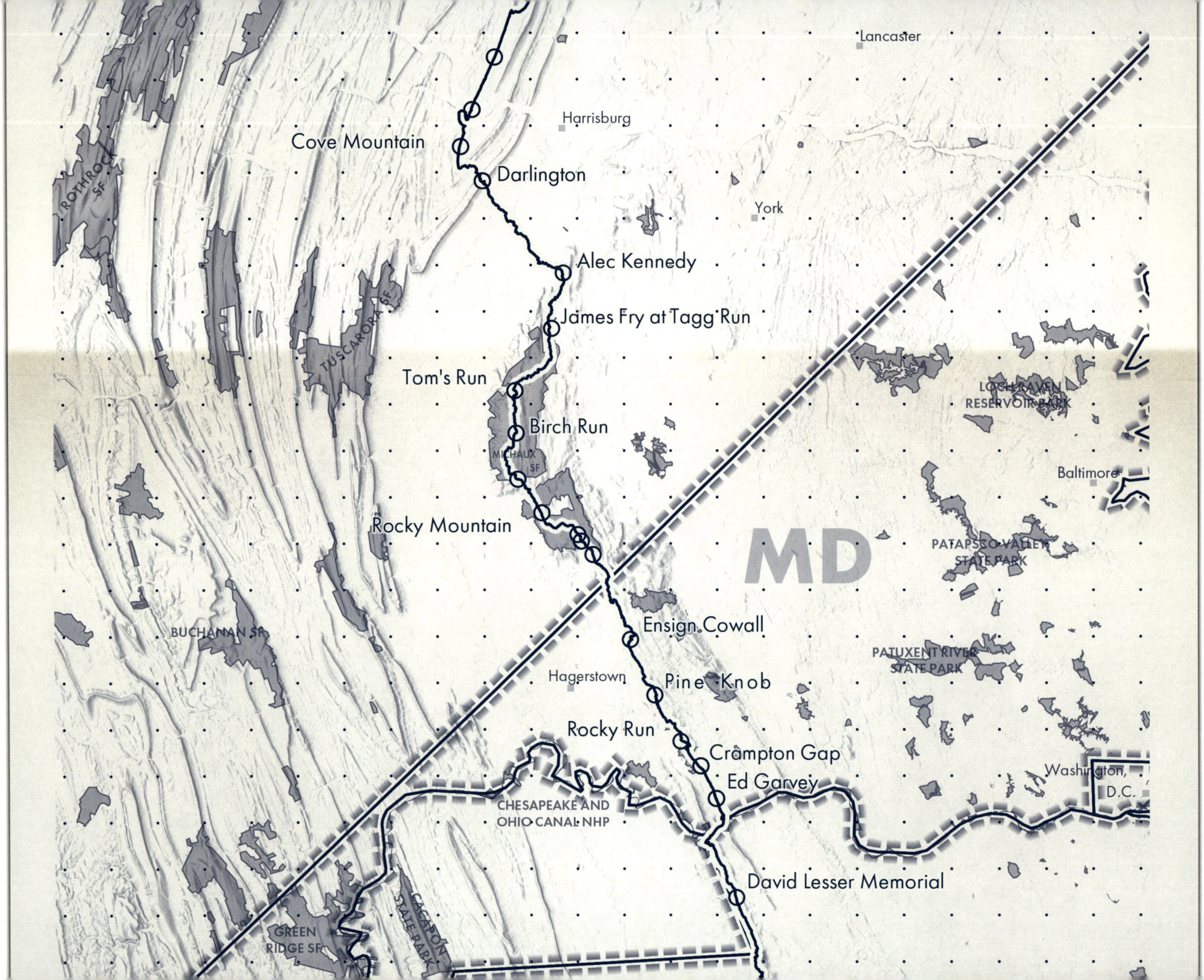

Lancaster
Harrisburg
Cove Mountain
Darlington
ROTHROCK SF
York
Alec Kennedy
James Fry at Tagg Run
TUSCARORA SF
Tom's Run
LOCH RAVEN RESERVOIR PARK
Birch Run
MICHAUX SF
Baltimore
Rocky Mountain
MD
PATAPSCO VALLEY STATE PARK
Ensign Cowall
BUCHANAN SF
PATUXENT RIVER STATE PARK
Hagerstown
Pine Knob
Rocky Run
Crampton Gap
Ed Garvey
Washington, D.C.
CHESAPEAKE AND OHIO CANAL NHP
David Lesser Memorial
CACAPON STATE PARK
GREEN RIDGE SF

« Appalachian Trail Fold Out Shelter Reference and Journal-Atop Adventure Map

John Nelson and Erich Rainville

Several years ago I was given a set of Field Notes brand notebooks by my colleague, Erich Rainville. It was a lovely collection of three, each showing a section of coastal contour lines; when the three notebooks were positioned vertically, the coverage spanned the West Coast of the United States. Some time later, Erich and I were chatting about the charm and materiality of the Field Notes aesthetic, and the conversation migrated to our shared appreciation of hiking journals: those tattered and stained treasures bursting with landscape sketches, pressed leaves, stickers, and the densely penned personal experience of the adventurer flowing through each page. Absolutely works of art.

We decided to design a rather minimal pocket map informed by the aesthetics of Field Notes and the expectant pages of a journal, awaiting the pen of a traveler—not an actual printed map, but an exercise in conjuring a wished-for product.

The result is a fully digital incarnation of something that appears convincingly physical: a journal map of the Appalachian Trail (fortuitously long and straight) with plenty of open space for a traveler's note-taking. It was a blast—virtually—embossing the bold Futura font into Kraft cardstock covers complete with pressed printing artifacts on the reverse, giving the blue "ink" a soaked and bled appearance, and adding a grommet hole for no good reason at all.

The map itself is limited to a single hue to echo the design clarity, and inexpensive production, of one-spot color printing, and the content is intentionally spartan, to invite the penned adventures of a would-be owner. Shelters are labeled throughout the trail, and a minimal sense of context is provided by state borders, muted topographic hillshade, and park lands. A 5 × 5-mile grid doubles as a scale hack and an homage to the grid-ruled pages of pocket memo books.

We were inspired by long maps, including medieval pilgrimage maps, pre-internet road trip "triptiks," a kitschy, cartoon midcentury industrial manual, an accordion-folded angler's map of the nineteenth century, and more recently the long river atlas project of Travis White. One of the wonders of the creative process is to find oneself steeped in the beautiful work of designers and cartographers of the past and today, and to be surprised by what comes of it.

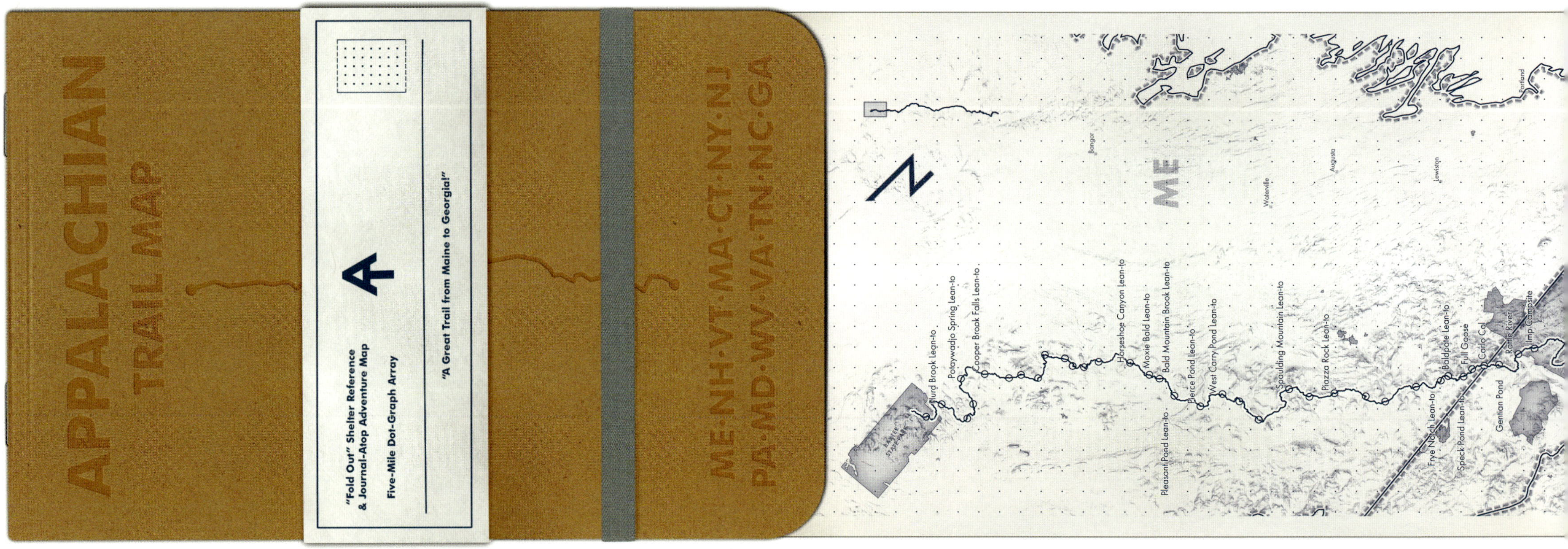
APPALACHIAN
TRAIL MAP
“Fold Out” Shelter Reference & Journal-Atop Adventure Map
Five-Mile Dot-Graph Array
“A Great Trail from Maine to Georgia!”
ME·NH·VT·MA·CT·NY·NJ
PA·MD·WV·VA·TN·NC·GA
ME
Bangor
Waterville
Augusta
Lewiston
Portland
Hurd Brook Lean-to
Potaywadjo Spring Lean-to
Cooper Brook Falls Lean-to
Horseshoe Canyon Lean-to
Moxie Bald Lean-to
Bald Mountain Brook Lean-to
Pleasant Pond Lean-to
Pierce Pond Lean-to
West Carry Pond Lean-to
Spaulding Mountain Lean-to
Piazza Rock Lean-to
Frye Notch Lean-to
Baldpate Lean-to
Speck Pond Lean-to
Full Goose
Carlo Col
Gentian Pond
Rattle River
Imp Campsite

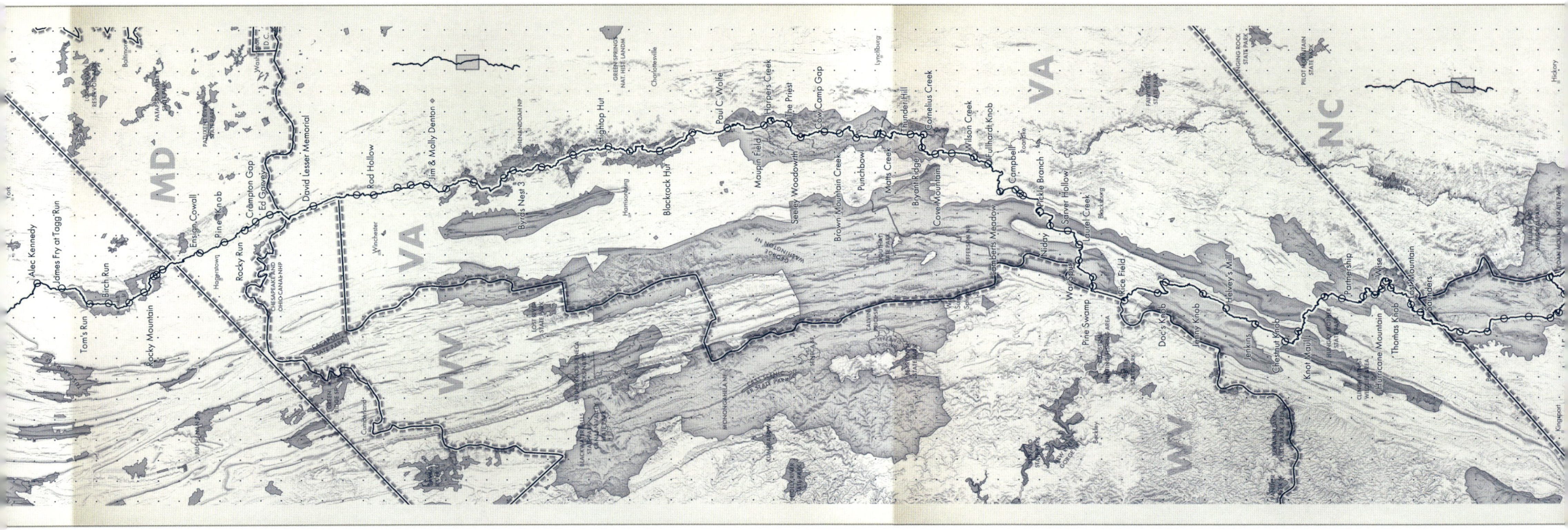
MD
VA
WV
VA
WV
NC
Alec Kennedy
James Fry at Tagg Run
Tom's Run
Birch Run
Rocky Mountain
Ensign Cowall
Pine Knob
Rocky Run
Crampton Gap
Ed Garvey
David Lesser Memorial
Rod Hollow
Jim & Molly Denton
Byrds Nest 3
Lighthouse Hut
Blackrock Hut
Paul C. Wolfe
Maupin Field
Harpers Creek
The Priest
Seeley-Woodworth
Cow Camp Gap
Brown Mountain Creek
Punchbowl
Matts Creek
Thunder Hill
Bryant Ridge
Cornelius Creek
Cove Mountain
Wilson Creek
Fullhardt Knob
Lamberts Meadow
Campbell
Pickle Branch
Sarver Hollow
Laurel Creek
Niday
Wapiti
Pine Swamp
Rice Field
Doc's Knob
Jenny Knob
Helveys Mill
Jenkins
Chestnut Knob
Knot Maul
Hurricane Mountain
Thomas Knob
Partnership
Wise
Lost Mountain
Saunders
Lynchburg
Charlottesville
Harrisonburg
Winchester
Hagerstown
Cumberland
Roanoke
Blacksburg
Hickory
Kingsport

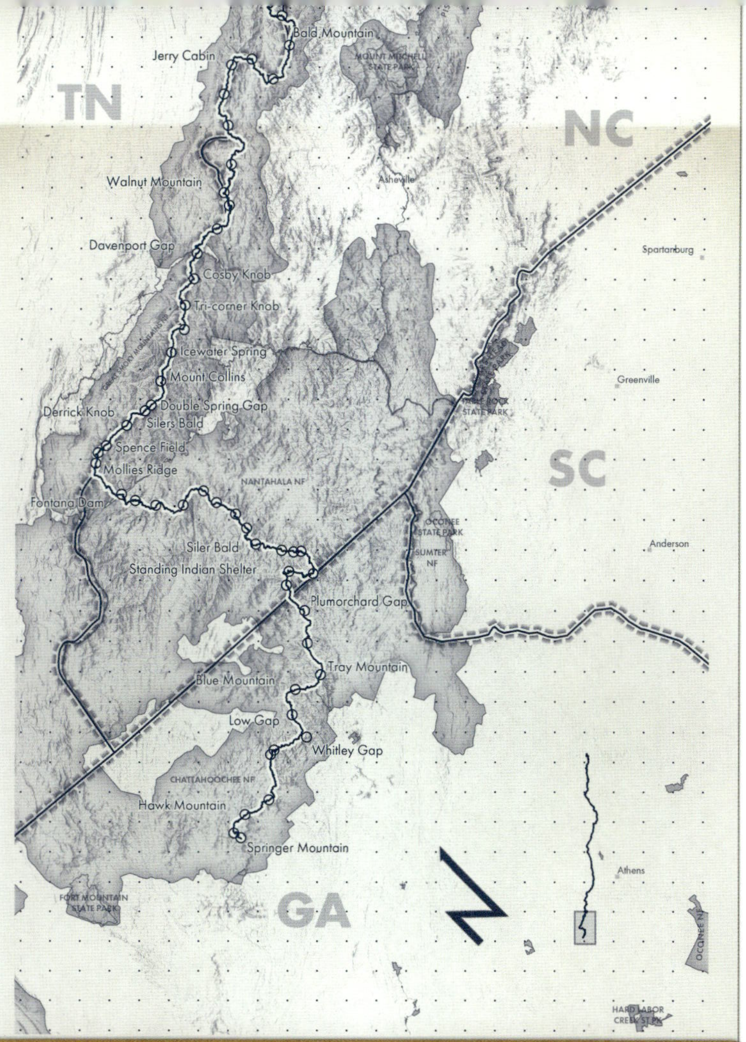

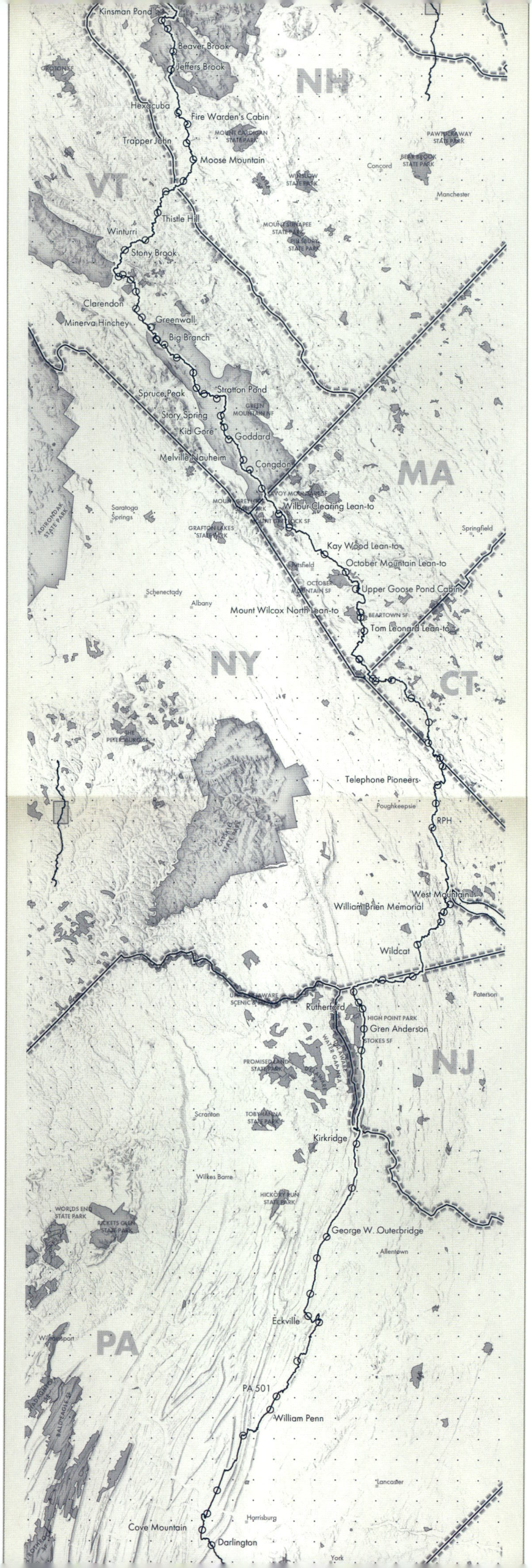

THIS MAP IS PROPERTY OF

NOTES ON GEOGRAPHIC SCALE

7 MILES / 5 MILES / 5 MILES

This map is equipped with a dot-graph array of 5 miles, as the bird flies. According to Pythagoras, diagonal distances between dots are just over 7 miles.
The winding nature of the AT implies a much greater hiked distance, however. The full journey is 2,178 actually-hiked miles, as you surely know well.

NOTABLE DATES

Begun .... Pine Grove Furnace .... Completed ....

Georgia	New Jersey ....
North Carolina	New York ....
Tennessee	Connecticut ....
Virginia	Massachusetts ....
West Virginia	Vermont ....
Maryland	New Hampshire ....
Pennsylvania	Maine ....

MAP SOURCES

AT Trail & Shelters	National Park Service
State Borders	Natural Earth
State & National Parks & Forests	Esri, NPS, TomTom
Terrain	Esri Living Atlas (Airbus, USGS, NGA NASA, CGIAR, NLS OS, NMA, Geodatastyrelsen, GSA, GSI)

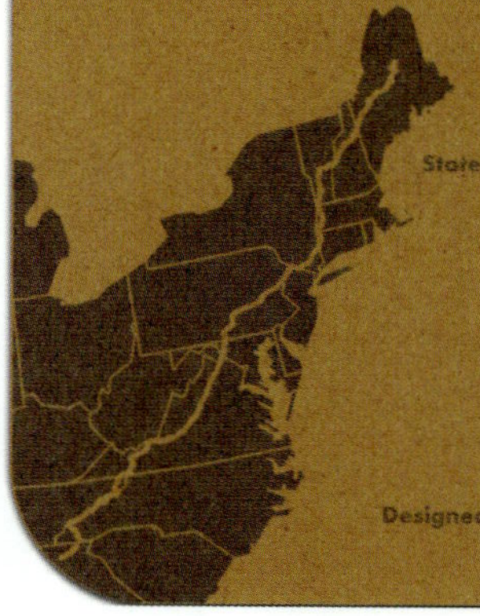

Designed by John Nelson & Erich Rainville

Cross-sections through California

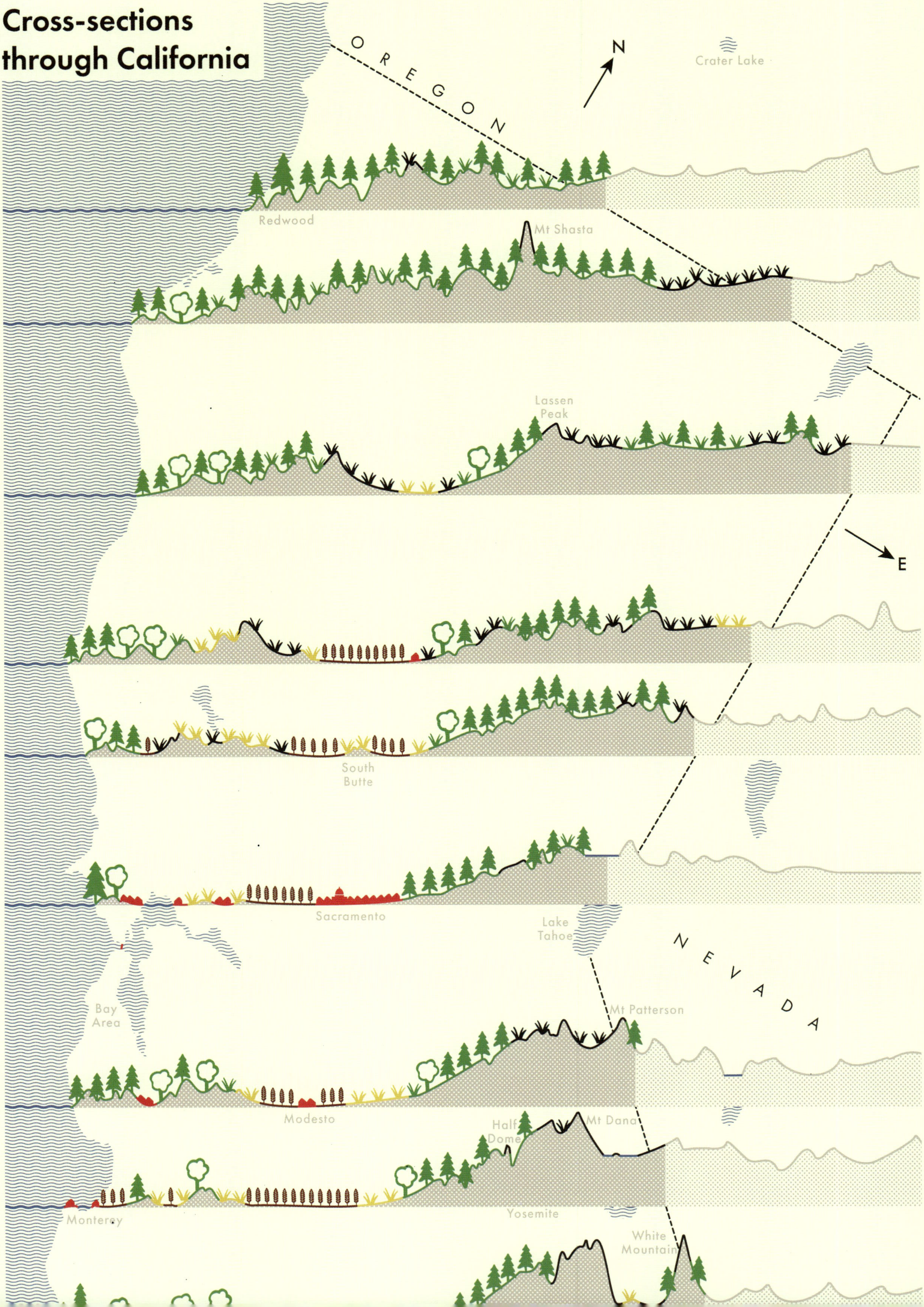

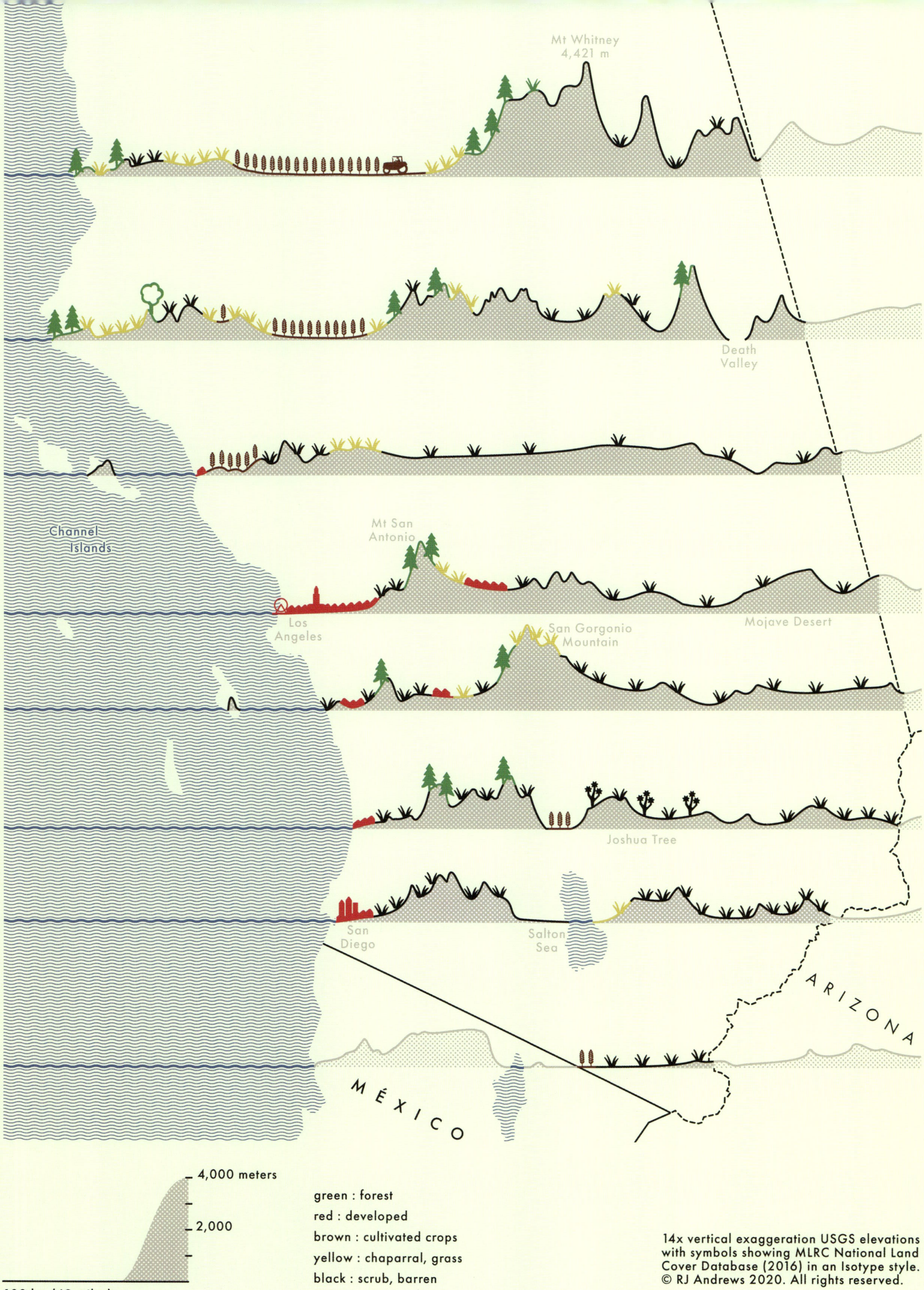

14x vertical exaggeration USGS elevations with symbols showing MLRC National Land Cover Database (2016) in an Isotype style.

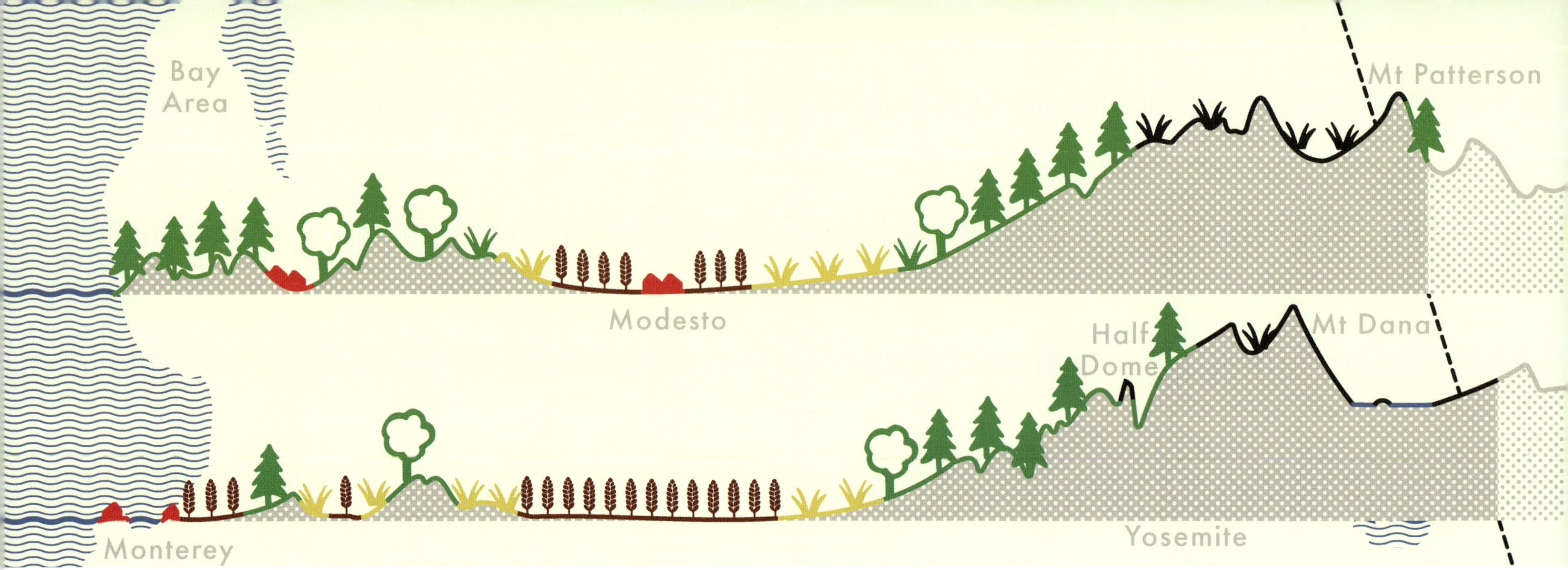

« Cross-sections through California

RJ Andrews

The purpose of this colorful map is to help the reader visualize California's vast landscape. A broad and general impression is emphasized over any specific details. The map takes inspiration from an information-graphic teaching system associated with the Isotype Institute. Its methodology was developed by Otto Neurath, Marie (Reidemeister) Neurath, and Gerd Arntz. Isotype design focuses attention on the essential facts by aggregating data into simple forms. Imitating Isotype graphics reveals how sophisticated the construction is—a property concealed by the approachable and memorable appearance. Otto Neurath wrote that "the purpose of teaching-pictures is to have an effect on the mind." It is my hope that this view of California helps you learn a little more about this fascinating state.

California's terrain is rich with wonders. Its geographic character is illustrated with elevation profiles and land-cover symbols. The map abstracts its input elevation data into a caricature by showing specific profiles and smoothed terrain. Profiles are chosen to feature famous landmarks like Yosemite and Death Valley. Together, the profiles suggest large features like the Central Valley and Sierra Nevada mountain range. Symbols represent what the land is like along each profile, using bold colors. They show transitions between temperate and arid climates. The symbols also show divisions between domesticated and wild regions.

North points away from the common north-is-up orientation. This rotation fits the state shape to the rectangular sheet. It also helps us to see California as we imagine it: the valley and mountains go up and down. The coast faces us outward, toward the Pacific Ocean.

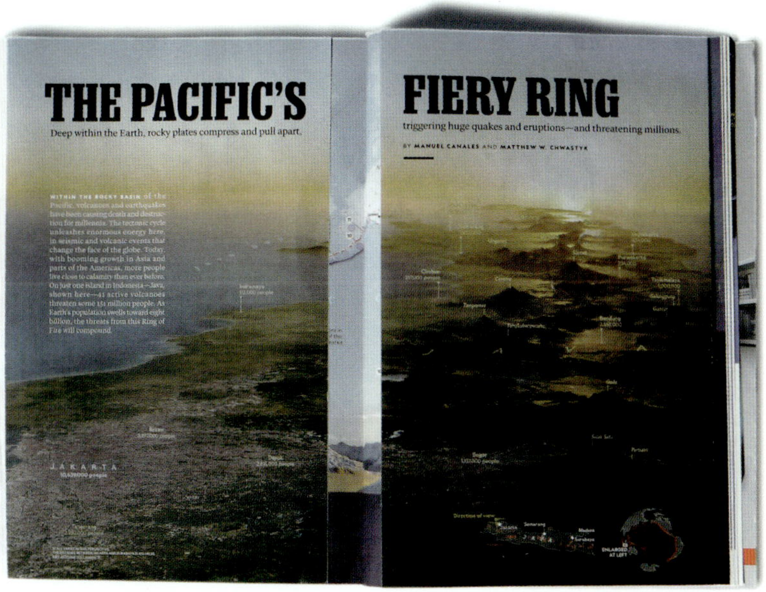

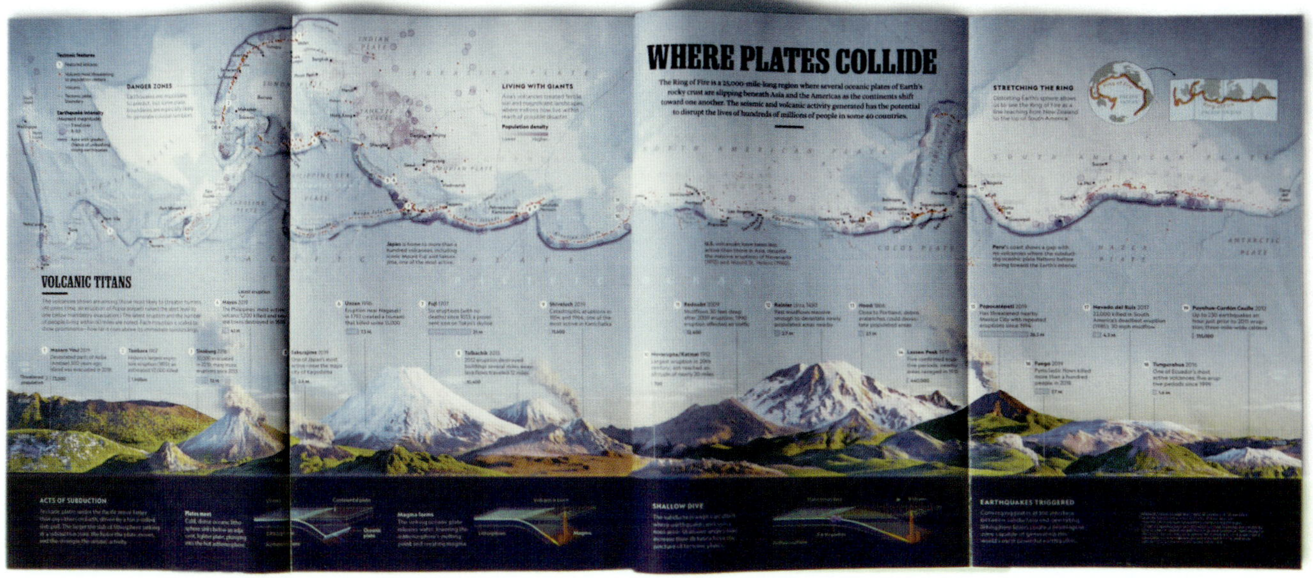

The Pacific's Fiery Ring »

Matthew W. Chwastyk and Manuel Canales — National Geographic

The most active volcanic areas in the world ring the Pacific, giving it the moniker "Ring of Fire." Manuel and I wanted to take a new and unique look at this region. To introduce the concept of people making a living in the shadow of volcanism, we opened this graphic feature with a perspective view of Java, where 151 million people share the island with 41 active volcanoes.

Moving into the interior of these double gatefolds, we worked with Charles Preppernau to develop a completely new look at the Ring of Fire by creating a new map that "unrolls" the coastline, allowing us to stretch out the volcanic regions and show threatening volcanoes and the places most likely to experience large earthquakes. This is revealed along the top of the pages and gradually fades from bathymetry to a false horizon—a hypothetical range of peaks, the most dangerous volcanoes on the Ring. We used digital elevation models and photography of the peaks to place in common scale all of the 19 volcanoes most threatening to human populations. Scientists combed through historical records, recent studies, and the present state of activity of the peaks in order to arrive at this list.

At the bottom of the pages, we illustrate the processes that lead to volcano formation and show why some areas tend to express tectonic activity as earthquakes more than volcanoes. The angle of subduction, where one rocky plate is diving beneath another, contributes to that distinction. We were inspired by many hand-drawn examples from the past two centuries, which usually presented a simplified view of peaks. One in particular—*Heights of the Principal Mountains in the World. Lengths of the Principal Rivers in the World*, by S. Augustus Mitchell, published in 1846—served as the impetus to pursue the peaks in scale. Manuel and I took a similar approach, but using real data and imagery, to create a graphic where the real appearance and scale of these volcanoes could be compared.

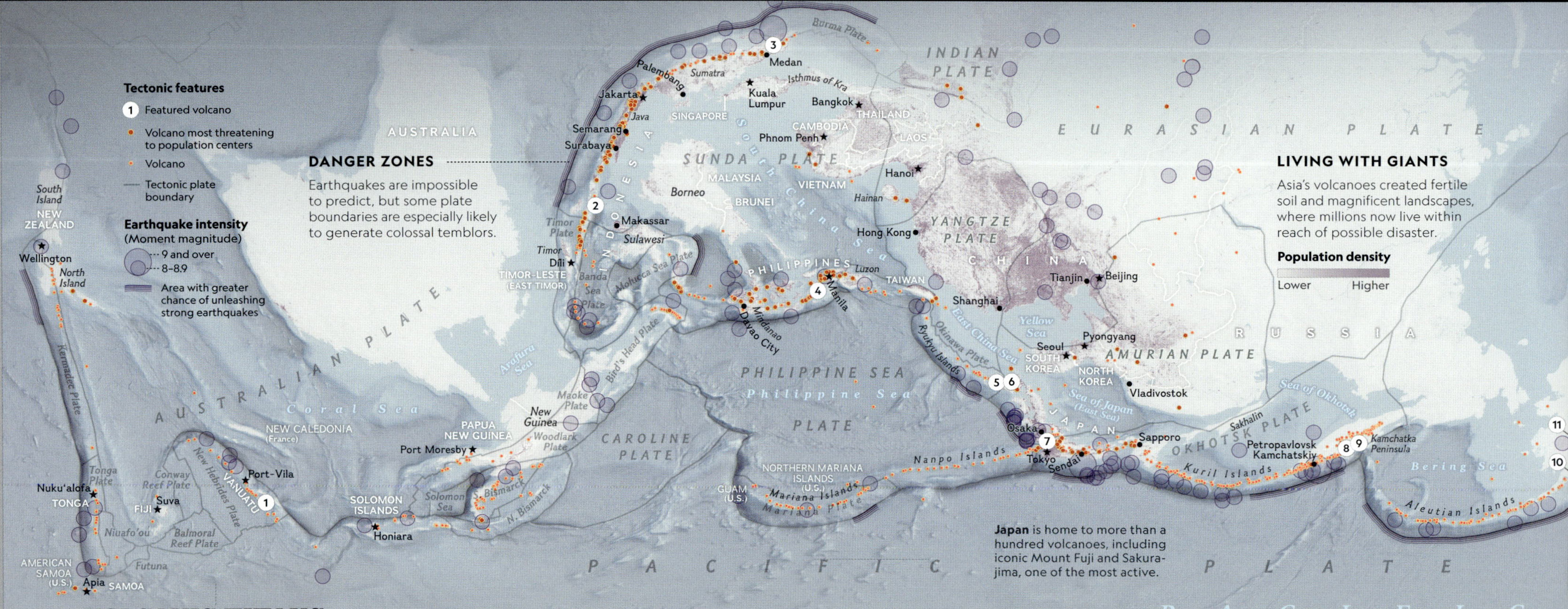

VOLCANIC TITANS

The volcanoes shown are among those most likely to threaten humans. (At press time, an eruption of Popocatépetl raised the alert level to one below mandatory evacuation.) The latest eruption and the number of people living within 60 miles are noted. Each mountain is scaled to show prominence—how far it rises above its immediate surroundings.

Latest eruption

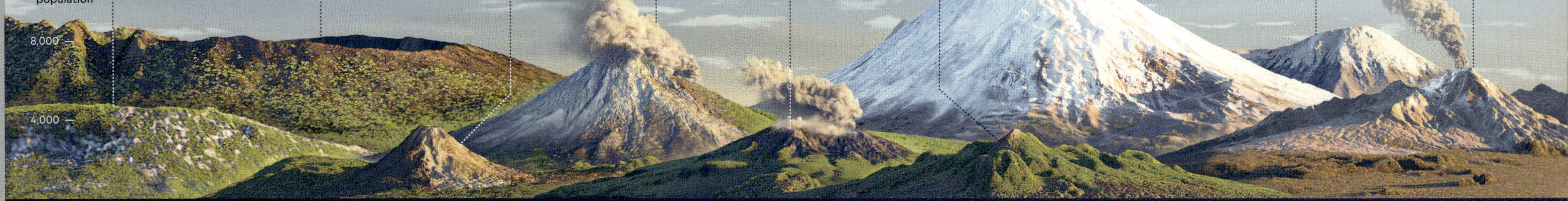

ACTS OF SUBDUCTION

Tectonic plates under the Pacific move faster than any others on Earth, driven by a force called slab pull. The larger the slab of lithosphere sinking at a subduction zone, the faster the plate moves, and the stronger the seismic activity.

Plates meet
Cold, dense oceanic lithosphere sinks below an adjacent, lighter plate, plunging into the hot asthenosphere.

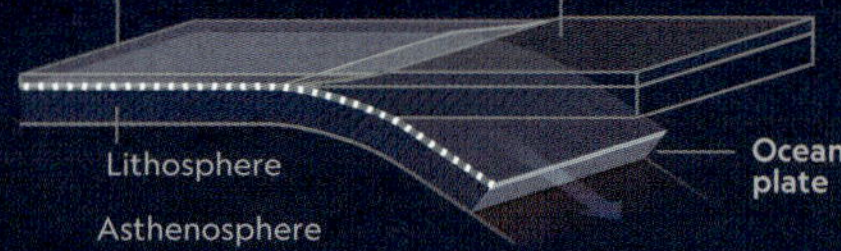

Magma forms
The sinking oceanic plate releases water, lowering the asthenosphere's melting point and creating magma.

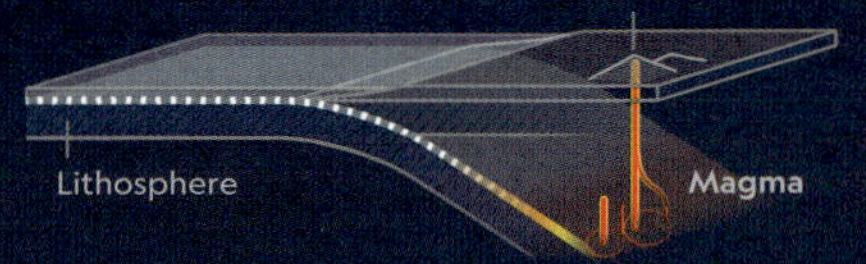

WHERE PLATES COLLIDE

The Ring of Fire is a 25,000-mile-long region where several oceanic plates of Earth's rocky crust are slipping beneath Asia and the Americas as the continents shift toward one another. The seismic and volcanic activity generated has the potential to disrupt the lives of hundreds of millions of people in some 40 countries.

STRETCHING THE RING

Distorting Earth's sphere allows us to see the Ring of Fire as a line reaching from New Zealand to the tip of South America.

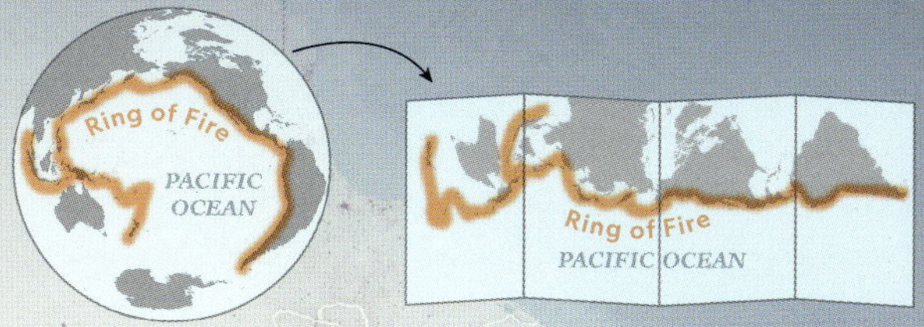

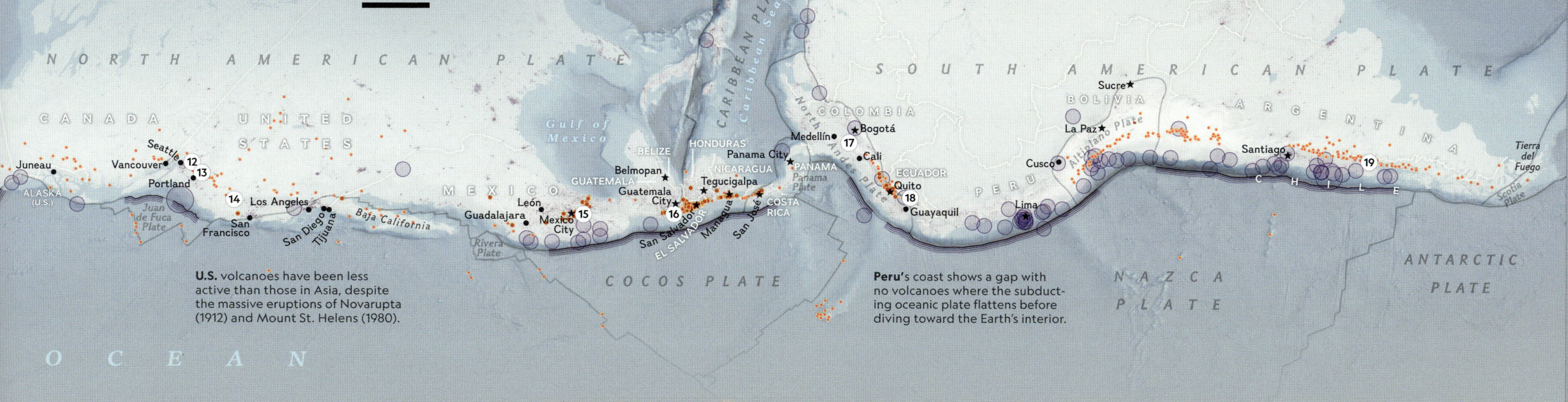

11 Redoubt 2009
Mudflows 30 feet deep after 2009 eruption; 1990 eruption affected air traffic
32,400

12 Rainier circa 1450
Past mudflows massive enough to devastate newly populated areas nearby
2.7 M

13 Hood 1866
Close to Portland; debris avalanches could devastate populated areas
2.1 M

15 Popocatépetl 2019
Has threatened nearby Mexico City with repeated eruptions since 1994
26.5 M

17 Nevado del Ruiz 2017
23,000 killed in South America's deadliest eruption (1985); 30 mph mudflow
4.3 M

19 Puyehue-Cordón Caulle 2012
Up to 230 earthquakes an hour just prior to 2011 eruption; three-mile-wide caldera
515,000

10 Novarupta/Katmai 1912
Largest eruption in 20th century; ash reached an altitude of nearly 20 miles
700

14 Lassen Peak 1917
Five confirmed eruptive periods; nearby areas ravaged in 1915
440,000

16 Fuego 2019
Pyroclastic flows killed more than a hundred people in 2018
7.7 M

18 Tungurahua 2016
One of Ecuador's most active volcanoes; five eruptive periods since 1999
1.6 M

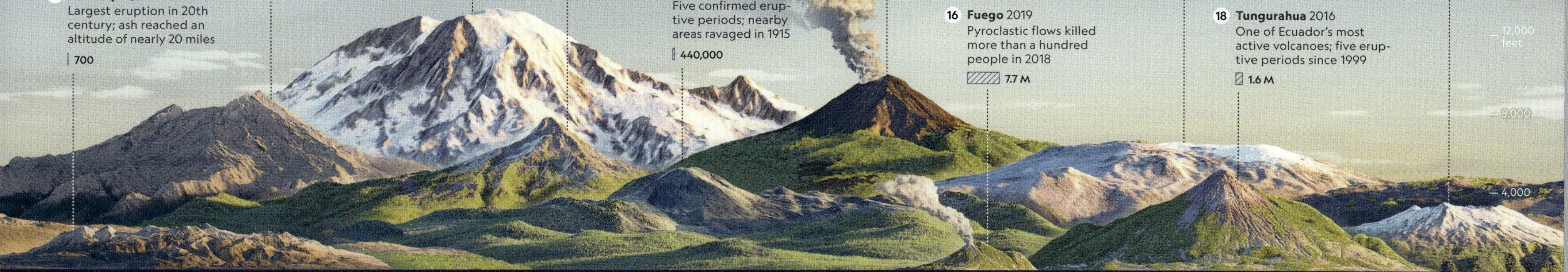

SHALLOW DIVE

The subduction angle can affect where earthquakes and volcanoes arise: Shallower angles may increase their distance from the juncture of tectonic plates.

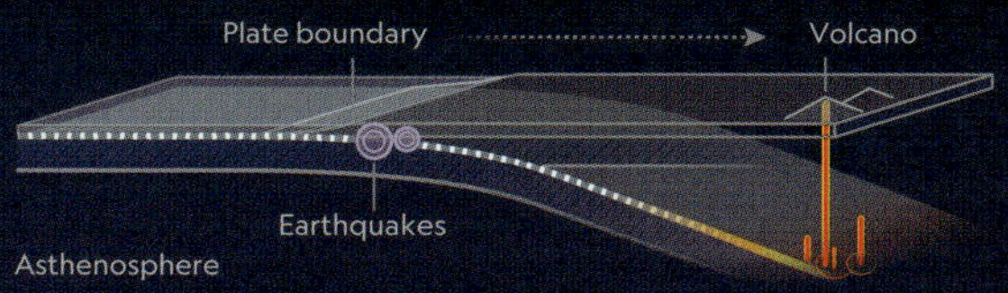

EARTHQUAKES TRIGGERED

Converging plates at the interface between subducting and overriding lithosphere layers create a seismogenic zone capable of generating the world's most powerful earthquakes.

MANUEL CANALES AND MATTHEW W. CHWASTYK, NGM STAFF; SEAN McNAUGHTON. VOLCANO ART: TOMÁŠ MÜLLER. MAP PROJECTION AND RENDERING: CHARLES PREPPERNAU
SOURCES: ADAM SOULE, WOODS HOLE OCEANOGRAPHIC INSTITUTION; ERIK KLEMETTI, DENISON UNIVERSITY; JANINE KRIPPNER, SMITHSONIAN INSTITUTION'S GLOBAL VOLCANISM PROGRAM; W.P. SCHELLART, VRIJE UNIVERSITEIT AMSTERDAM; GEOLOGICAL SOCIETY OF AMERICA; UN OFFICE FOR DISASTER RISK REDUCTION; USGS

I Was There »

Levi Westerveld and Anne Kelly Knowles

I Was There maps the experiences of Holocaust survivors Anna Patipa and Jacob Brodman, as told in videotaped interviews.

This map was born of necessity. Conventional cartography and GIS use latitude and longitude to locate places in space. Coordinates, however, are often impossible to determine from survivors' traumatic narratives, especially for small places that were central to their experiences.

We built the map by arranging large, known places, such as cities or concentration camps, according to their approximate geographic location. We represented smaller places, such as barracks, woods, or rooms, by their topological relationship—inside, outside, adjacent, and so forth—to other places mentioned in the interviews.

To combine large and small places, we developed a hybrid approach that combines GIS with manual, relative placement. Levi Westerveld drew each place in charcoal by hand to give it a personal quality, and to remind readers of the presence of the cartographer as an interpreter of the testimony. Places were scaled loosely in relation to one another. Different levels of opacity indicate how frequently a place was mentioned. Faint arrows trace each speaker's sequential movement between places. Finally, quotes embed each survivor's voice in the map, representing their memories of events in distinct spaces.

Place-dense spaces, such as Anna's and Jacob's hometowns, stand in contrast to places of forced labor and captivity, where movement and experience were limited. The Auschwitz death camp is an exception, perhaps because of the long time spent there. Small, stand-alone places were often experienced in transit, such as bridges where bystanders dropped bread into open cattle cars. Some large places (Budapest, Theresienstadt) were included because a relative was sent there.

Our map does not pursue geographical precision; rather, it is a representation of the complex, subjective geographies expressed in Patipa's and Brodman's testimonies.

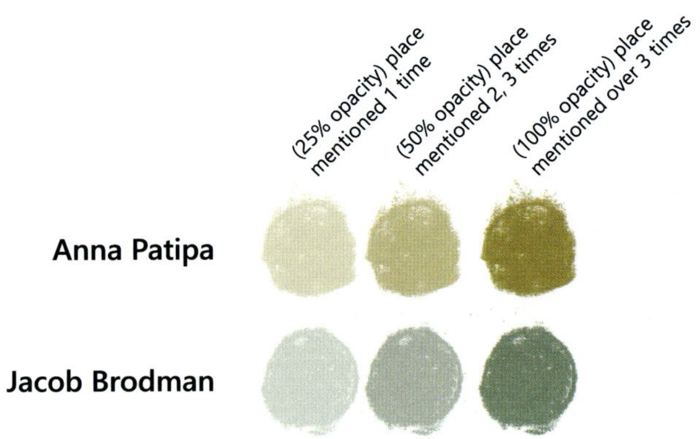

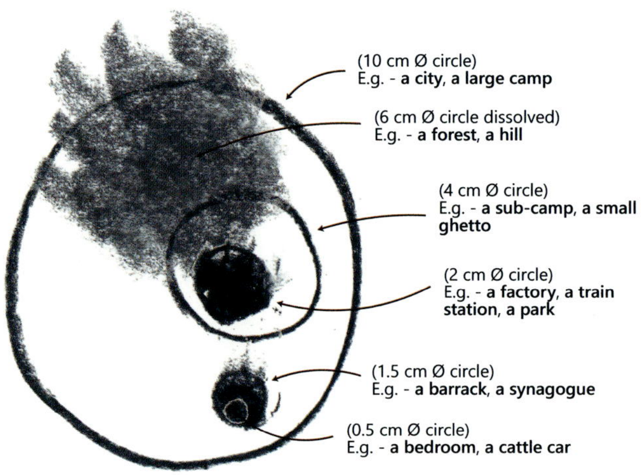

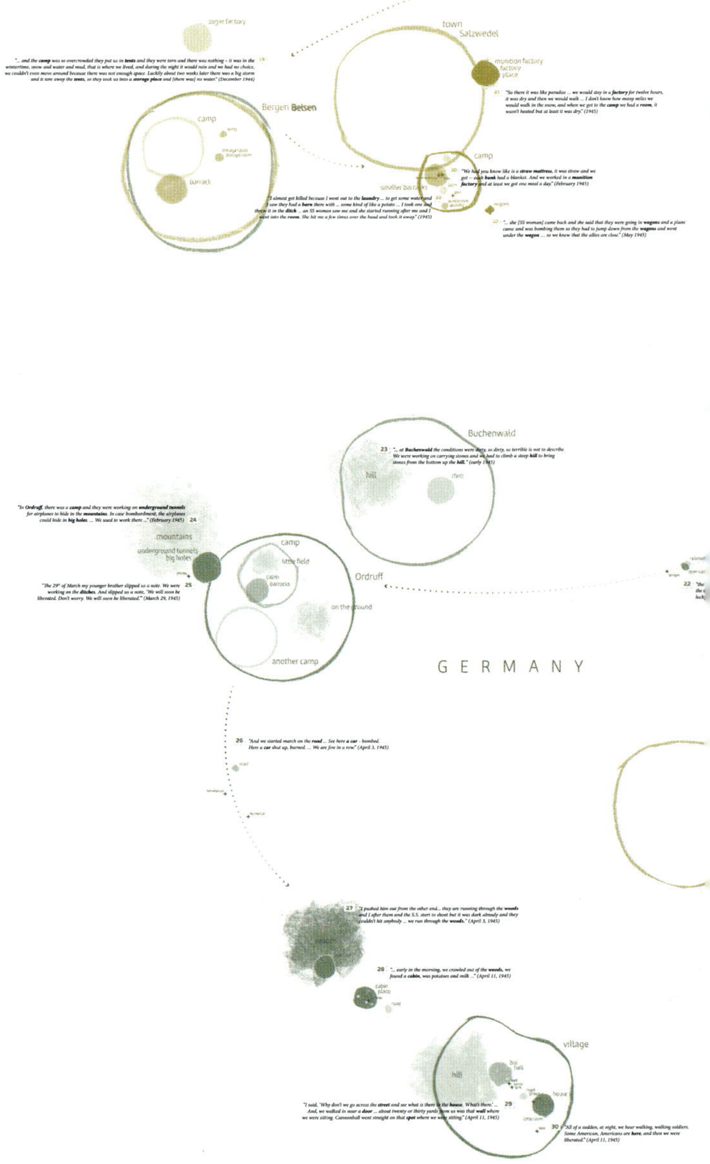

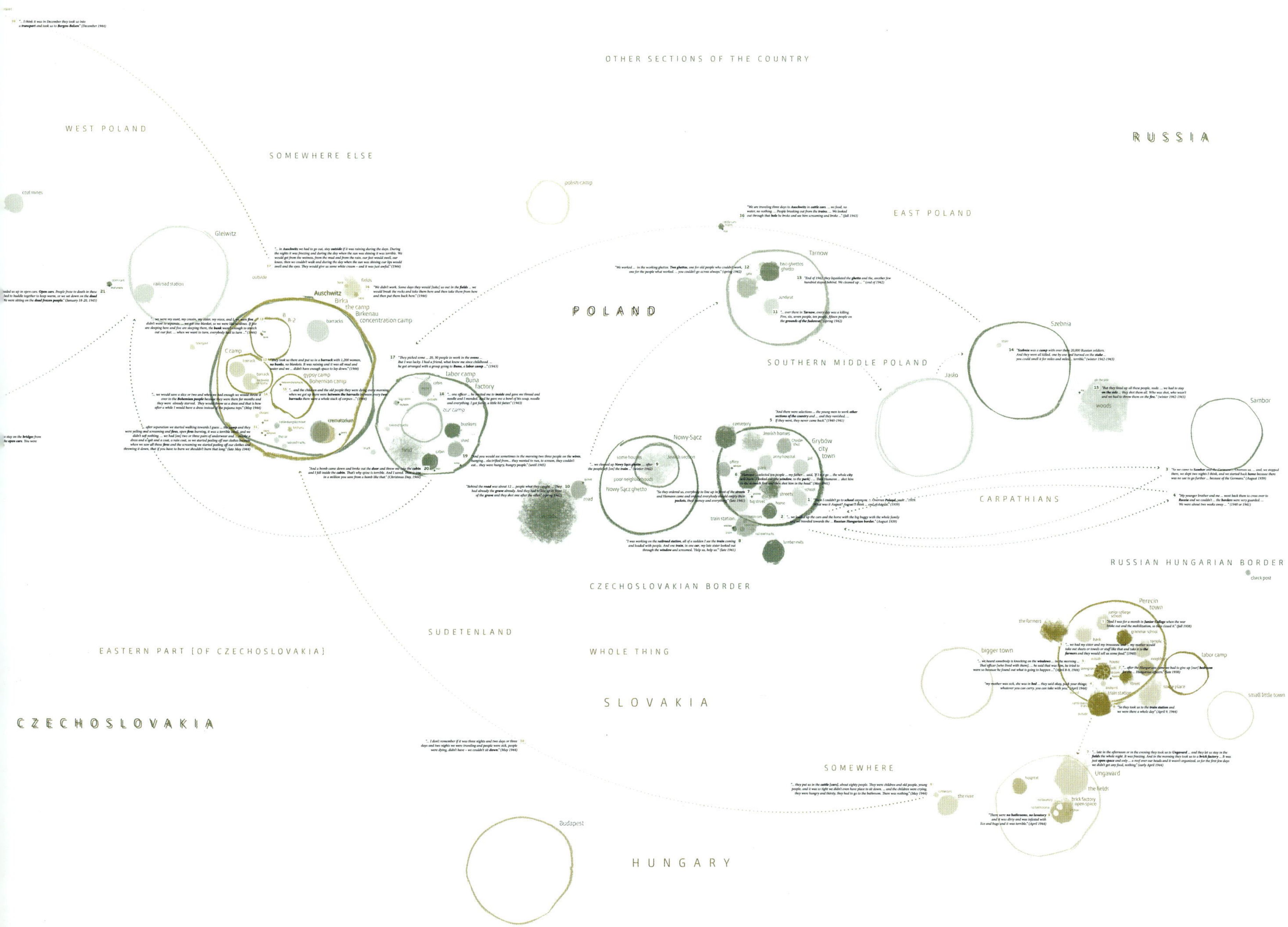

I Was There

PLACES OF EXPERIENCE IN THE HOLOCAUST

How to read this map

This map shows the many places mentioned in interviews with Anna Patipa and Jacob Brodman, two Holocaust survivors. You can follow the sequence of places in their accounts by noting the number of each excerpt, generally from east (right) to west (left). The largest places, such as cities or concentration camps, are located according to their approximate actual location. Smaller places, such as barracks, woods, or rooms, are located based on their topological relationship to the other places mentioned in the interviews, such as places they were within or near. The legend explains the colors that signify each person's account and the reasoning behind the map's style and effects. If both survivors mention a place, it appears in both colors.

Why was the map made and what does it say?

This map was born of necessity. Conventional cartography and GIS use latitude and longitude to place objects in space. However, coordinates are often impossible to determine for small places or for historical places that have ambiguous or uncertain location. Because we wanted to include all places that seemed significant to Patipa and Brodman, we developed a hybrid approach that enables us to include the very small, intimate places (a building, a train car, a pocket) as well as towns, regions, and countries. In this way the map is a cartographic exploration of the spatiality of the Holocaust as well as our own responses to the accounts. The textual excerpts on the map either describe the quality of a place or an event that made it notable for the speaker. The texts do not constitute narratives, but we hope they give you a glimpse of two individuals' experiences as the Holocaust tore them from their homes and carried them across Europe.

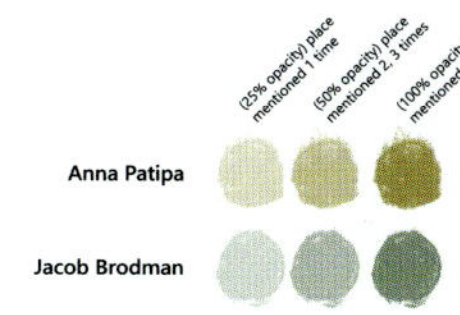

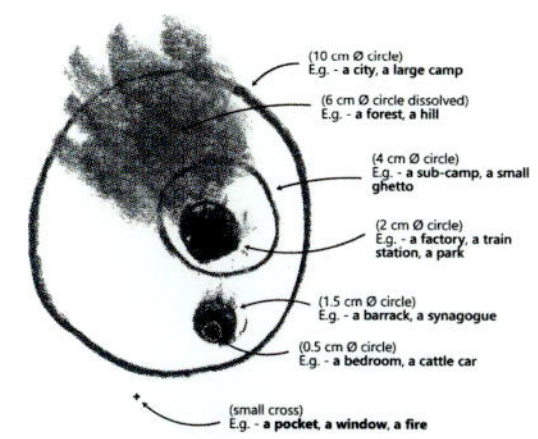

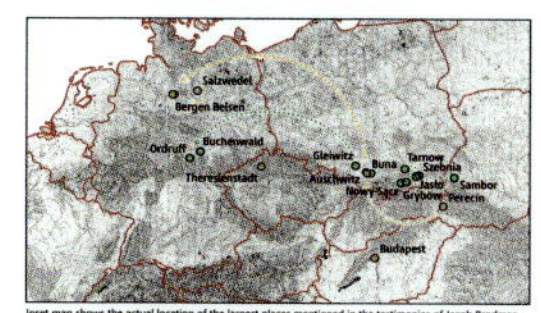

Inset map shows the actual location of the largest places mentioned in the testimonies of Jacob Brodman and Anna Patipa. This map uses circa 2017 country borders that are different from the changing European country borders from during the Holocaust period.

Text excerpts: Interview of Anna Patipa (February 23, 1989) is from the archives of the Tauber Holocaust Library of the Jewish Family and Children's Services Holocaust Center, used by permission. Interview of Jack Brodman (April 13, 1989 done by the National Council of Jewish Women Sarasota-Manatee Section, Holocaust Oral History Project, on April 13, 1989, USHMM Collection reference no. RG-50.154*0007, used by permission of the United States Holocaust Memorial Museum.

Map: Coordinate locations from Wikipedia and Google Earth. Historical regional locations from Paul Robert Magocsi, Historical Atlas of Central Europe, rev. ed. (Seattle: University of Washington Press, 2002) and the dataset of World War II era historical territorial boundaries developed by Michael De Groot for the Holocaust Geographies Collaborative; see Michael De Groot, "Building the New Order: 1938-1945," August 24, 2010, Stanford Spatial History Lab, available online at https://web.stanford.edu/group/spatialhistory/cgi-bin/site/pub.php?id=51.

For their critiques and suggestions, we thank Michael Alpert, Tim Cole, Alberto Giordano, Justus Hillebrand, Stephen J. Hornsby, Paul B. Jaskot, and particularly Margaret Pearce.

I was There. Places of Experience in the Holocaust.
Map produced by Levi Westerveld and Anne Kelly Knowles (2018).

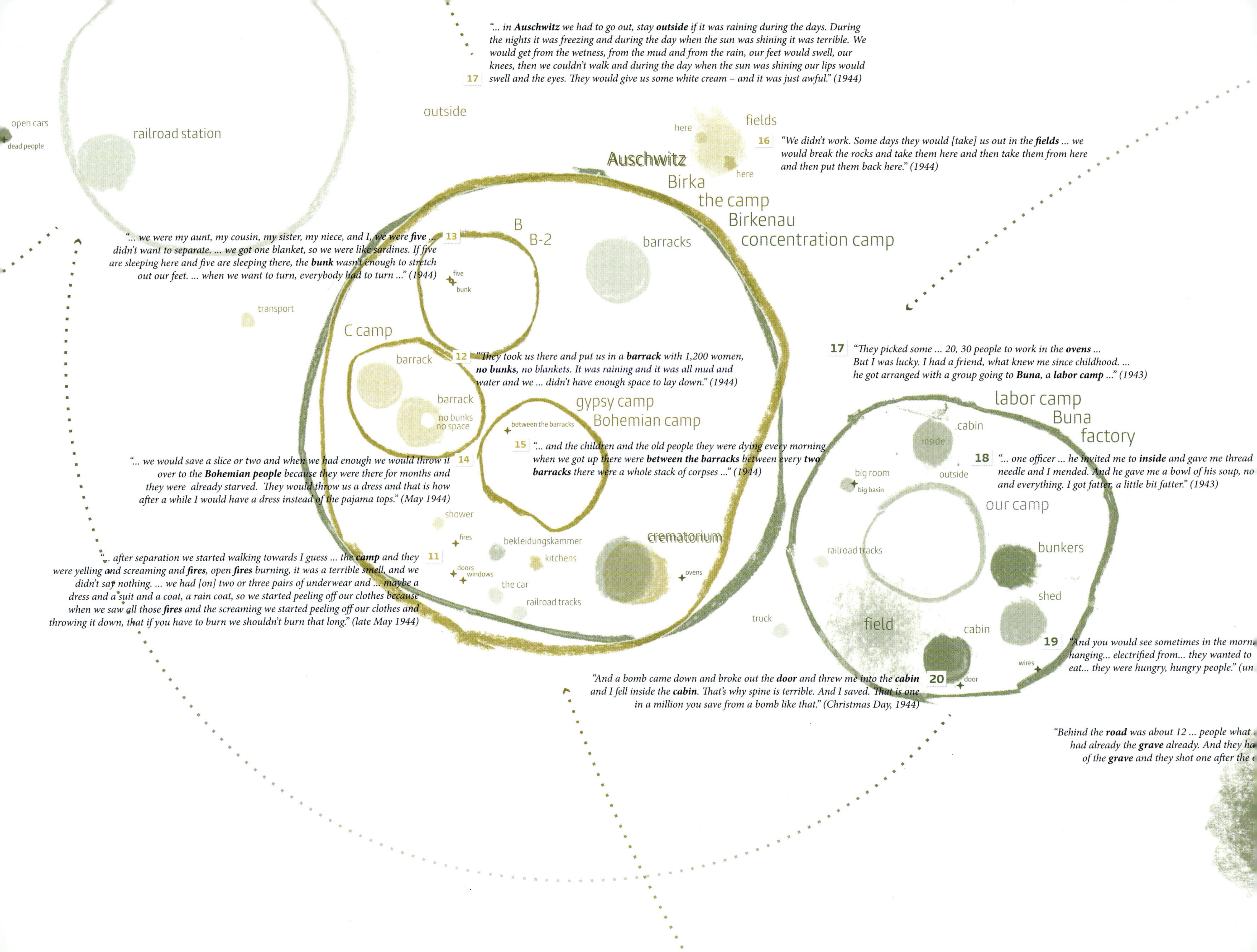

17 "... in **Auschwitz** we had to go out, stay **outside** if it was raining during the days. During the nights it was freezing and during the day when the sun was shining it was terrible. We would get from the wetness, from the mud and from the rain, our feet would swell, our knees, then we couldn't walk and during the day when the sun was shining our lips would swell and the eyes. They would give us some white cream – and it was just awful." (1944)

16 "We didn't work. Some days they would [take] us out in the **fields** ... we would break the rocks and take them here and then take them from here and then put them back here." (1944)

13 "... we were my aunt, my cousin, my sister, my niece, and I, we were **five** ... didn't want to separate. ... we got one blanket, so we were like sardines. If five are sleeping here and five are sleeping there, the **bunk** wasn't enough to stretch out our feet. ... when we want to turn, everybody had to turn ..." (1944)

12 "They took us there and put us in a **barrack** with 1,200 women, **no bunks**, no blankets. It was raining and it was all mud and water and we ... didn't have enough space to lay down." (1944)

17 "They picked some ... 20, 30 people to work in the **ovens** ... But I was lucky. I had a friend, what knew me since childhood. ... he got arranged with a group going to **Buna**, a **labor camp** ..." (1943)

15 "... and the children and the old people they were dying every morning when we got up there were **between the barracks** between every **two barracks** there were a whole stack of corpses ..." (1944)

14 "... we would save a slice or two and when we had enough we would throw it over to the **Bohemian people** because they were there for months and they were already starved. They would throw us a dress and that is how after a while I would have a dress instead of the pajama tops." (May 1944)

18 "... one officer ... he invited me to **inside** and gave me thread needle and I mended. And he gave me a bowl of his soup, no and everything. I got fatter, a little bit fatter." (1943)

11 "... after separation we started walking towards I guess ... the **camp** and they were yelling and screaming and **fires**, open **fires** burning, it was a terrible smell, and we didn't say nothing. ... we had [on] two or three pairs of underwear and ... maybe a dress and a suit and a coat, a rain coat, so we started peeling off our clothes because when we saw all those **fires** and the screaming we started peeling off our clothes and throwing it down, that if you have to burn we shouldn't burn that long." (late May 1944)

19 "And you would see sometimes in the morn hanging... electrified from... they wanted to eat... they were hungry, hungry people." (un

20 "And a bomb came down and broke out the **door** and threw me into the **cabin** and I fell inside the **cabin**. That's why spine is terrible. And I saved. That is one in a million you save from a bomb like that." (Christmas Day, 1944)

"Behind the **road** was about 12 ... people what had already the **grave** already. And they ha of the **grave** and they shot one after the

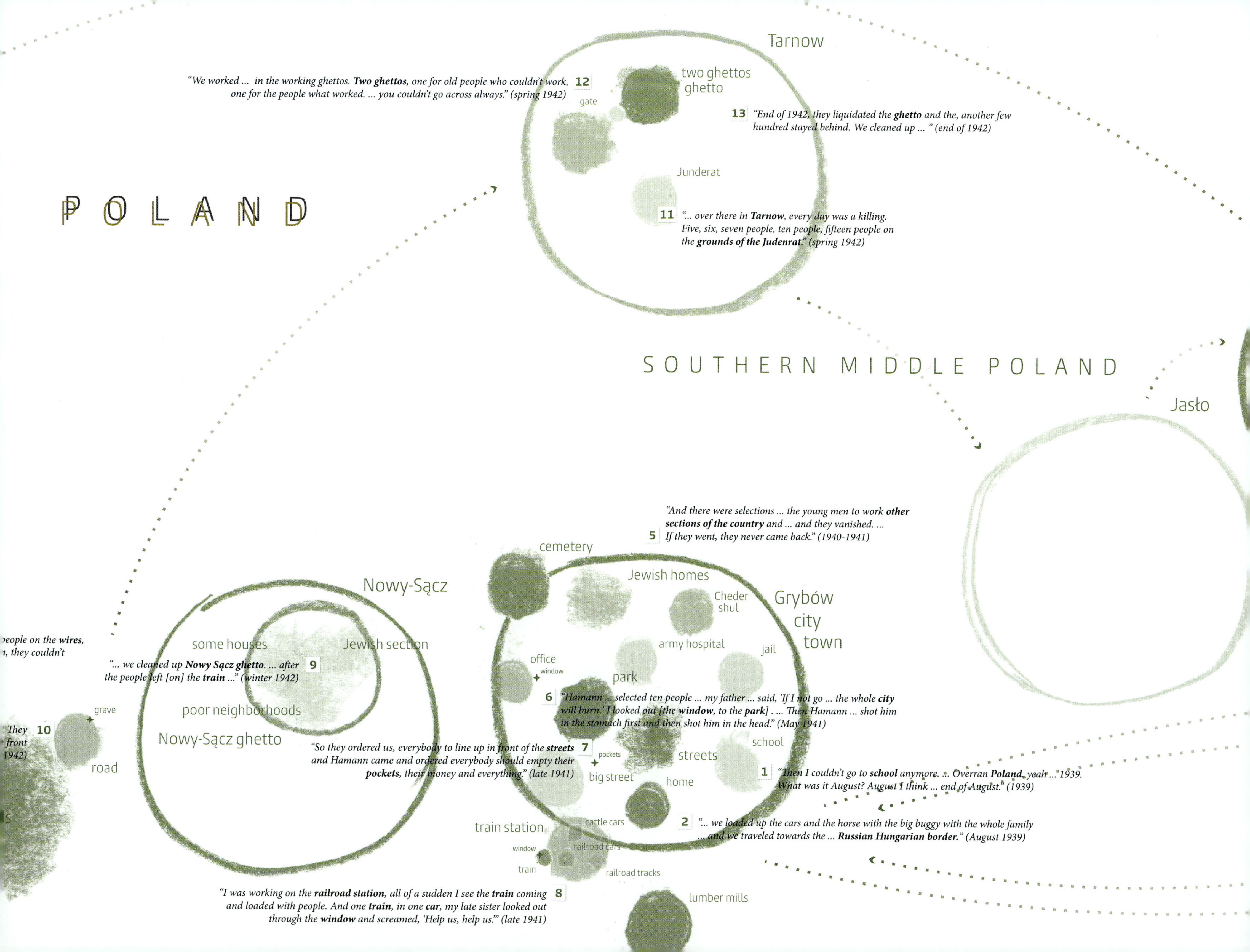

POLAND
Tarnow
two ghettos
ghetto
gate
12 "We worked ... in the working ghettos. **Two ghettos**, one for old people who couldn't work, one for the people what worked. ... you couldn't go across always." (spring 1942)
13 "End of 1942, they liquidated the **ghetto** and the, another few hundred stayed behind. We cleaned up ... " (end of 1942)
Junderat
11 "... over there in **Tarnow**, every day was a killing. Five, six, seven people, ten people, fifteen people on the **grounds of the Judenrat**." (spring 1942)
SOUTHERN MIDDLE POLAND
Jasło
5 "And there were selections ... the young men to work **other sections of the country** and ... and they vanished. ... If they went, they never came back." (1940-1941)
cemetery
Jewish homes
Cheder
shul
Grybów
city
town
army hospital
jail
office
window
park
6 "Hamann ... selected ten people ... my father ... said, 'If I not go ... the whole **city** will burn.' I looked out [the **window**, to the **park**] ... Then Hamann ... shot him in the stomach first and then shot him in the head." (May 1941)
school
7 "So they ordered us, everybody to line up in front of the **streets** and Hamann came and ordered everybody should empty their **pockets**, their money and everything." (late 1941)
pockets
streets
big street
home
1 "Then I couldn't go to **school** anymore. ... Overran **Poland**, yeah ... 1939. What was it August? August I think ... end of August." (1939)
2 "... we loaded up the cars and the horse with the big buggy with the whole family ... and we traveled towards the ... **Russian Hungarian border.**" (August 1939)
Nowy-Sącz
some houses
Jewish section
9 "... we cleaned up **Nowy Sącz ghetto**. ... after the people left [on] the **train** ..." (winter 1942)
poor neighborhoods
Nowy-Sącz ghetto
grave
10
road
train station
cattle cars
window
railroad cars
train
railroad tracks
8 "I was working on the **railroad station**, all of a sudden I see the **train** coming and loaded with people. And one **train**, in one **car**, my late sister looked out through the **window** and screamed, 'Help us, help us.'" (late 1941)
lumber mills

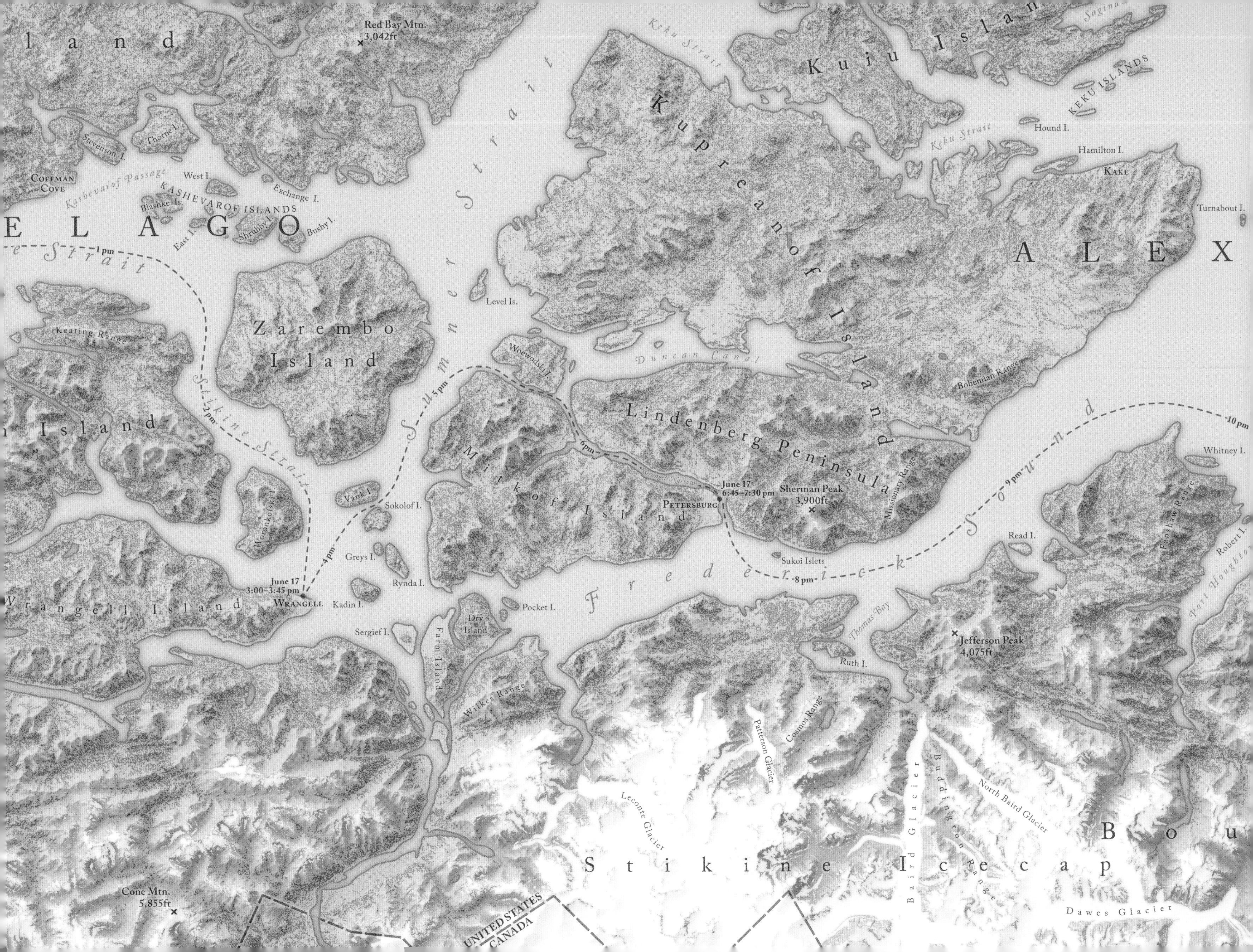

land
Red Bay Mtn.
3,042ft
Keku Strait
Kuiu Islan
KEKU ISLANDS
Hound I.
Keku Strait
Hamilton I.
Kake
Turnabout I.
Kupreanof Island
Sumner Strait
Thorne I.
Stevenson I.
Coffman Cove
Kashevarof Passage
West I.
Exchange I.
KASHEVAROF ISLANDS
Blashke Is.
Shrubby I.
Bushy I.
East I.
ELAGO
ALEX
e Strait
1pm
Level Is.
Zarembo Island
Keating Range
Woewodski I.
Duncan Canal
Bohemian Range
n Island
Stikine Strait
2pm
5pm
Lindenberg Peninsula
10pm
Whitney I.
6pm
Mitkof Island
June 17
6:45–7:30 pm
Petersburg
Sherman Peak
3,900ft
Missionary Range
9pm
Frederick Sound
Vank I.
Sokolof I.
Woronkofski I.
Fanshaw Range
Read I.
Robert I.
Greys I.
4pm
Sukoi Islets
June 17
3:00–3:45 pm
Wrangell
Rynda I.
8pm
Kadin I.
Wrangell Island
Pocket I.
Port Houghton
Dry Island
Sergief I.
Farm Island
Thomas Bay
Jefferson Peak
4,075ft
Ruth I.
Wilkes Range
Patterson Glacier
Cosmos Range
Leconte Glacier
Baird Glacier
Buddington Range
North Baird Glacier
Bou
Stikine Icecap
Cone Mtn.
5,855ft
UNITED STATES
CANADA
Dawes Glacier

> *I could look out at the landscape, point to this peak or that passage, and know one name that it might be called.*

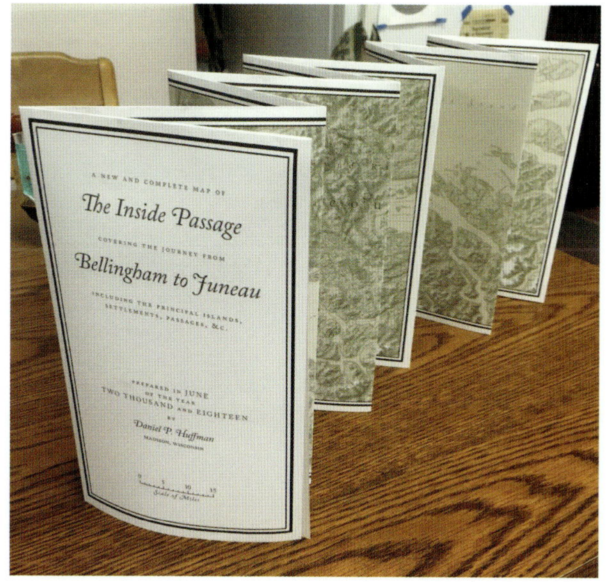

« A New and Complete Map of the Inside Passage, Covering the Journey from Bellingham to Juneau, Including the Principal Islands, Settlements, Passages, &c.

Daniel P. Huffman

In the summer of 2018, on my way to conduct a workshop in Alaska, I took a ferry through the Inside Passage. Before I left, I spent a couple of weeks making this double-sided, 10 × 60-inch folded map. The sepia-inspired colors, swash typography (I turn often to Adobe Caslon for this purpose), and lengthy title are a somewhat naive pastiche of older maps; I viewed the whole journey with a romantic sense of nautical adventure, even from the comfort of a heated vessel with a cafeteria (though I did have to sleep on the deck).

I made this map for two reasons. First, I wanted to be able to figure out where I was at any given time while on the trip. Since passengers were warned that cell reception would be unreliable during the journey, I decided a paper map was needed. To locate myself, I included a line for the ferry route, marked with estimates of what our position was likely to be at every hour along the journey, which I made by dividing the line into even segments based on the time we were scheduled to travel between each port and the next (this assumed that our speed remained constant).

The second, greater reason, though, was to achieve a familiarity with a new landscape. I knew little of the Inside Passage (or, really, the Pacific Northwest), and the slow, intentional process of cleaning up data, styling glaciers, carefully positioning labels, and so on gave me a chance to learn just a few of its curves and angles and words. I traveled the Inside Passage on my screen well before I set foot on the ferry, which made me feel a little less out of place in an unfamiliar land. As did knowing names: I could look out at the landscape, point to this peak or that passage, and know one name that it might be called. And though there are no *true* names for natural features, there was something that felt special and powerful in knowing some of them.

In the end, it turned out I could still use my phone's GPS and intermittent cell reception to follow my progress on Google Maps during my journey. However, I still pulled the paper map out quite often, both because it offered a more detailed view, and because I liked looking at it and enjoyed the memory of making it. But while it wasn't as *practically* useful as I had hoped, the knowledge I gained during the mapmaking process still made the journey far more satisfying.

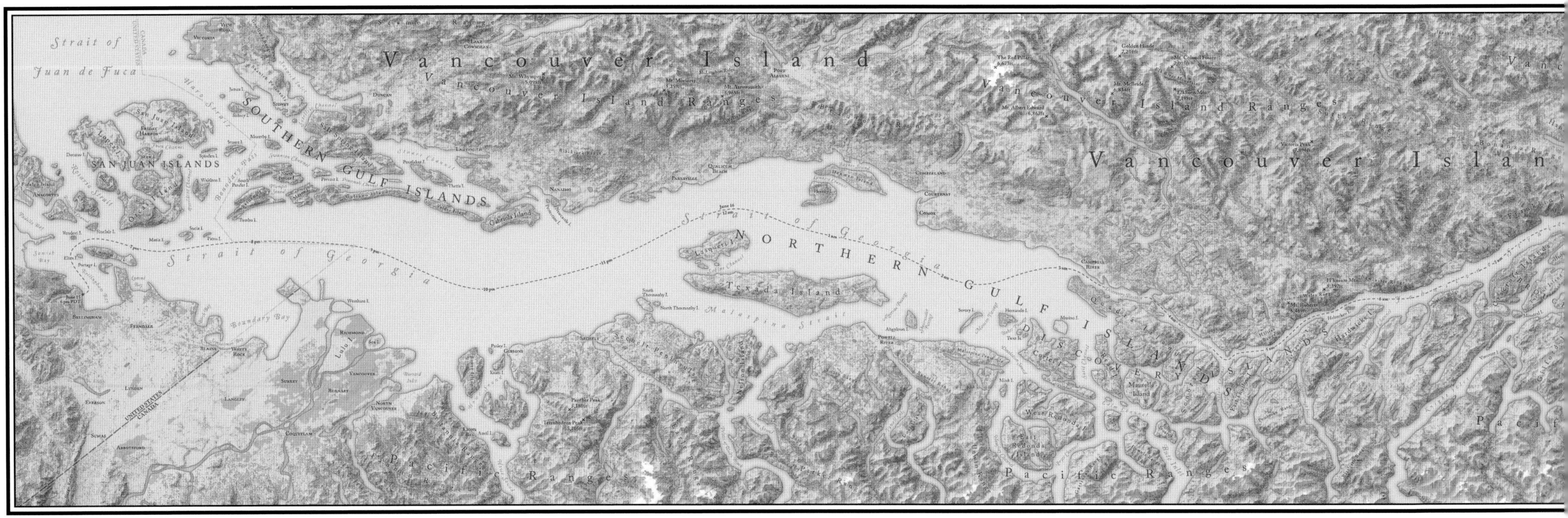
Strait of
Juan de Fuca
Vancouver Island
Vancouver Island Ranges
SOUTHERN GULF ISLANDS
SAN JUAN ISLANDS
Strait of Georgia
NORTHERN GULF ISLANDS
Texada Island
Malaspina Strait
Boundary Bay
Pacific Ranges
Vancouver Islan

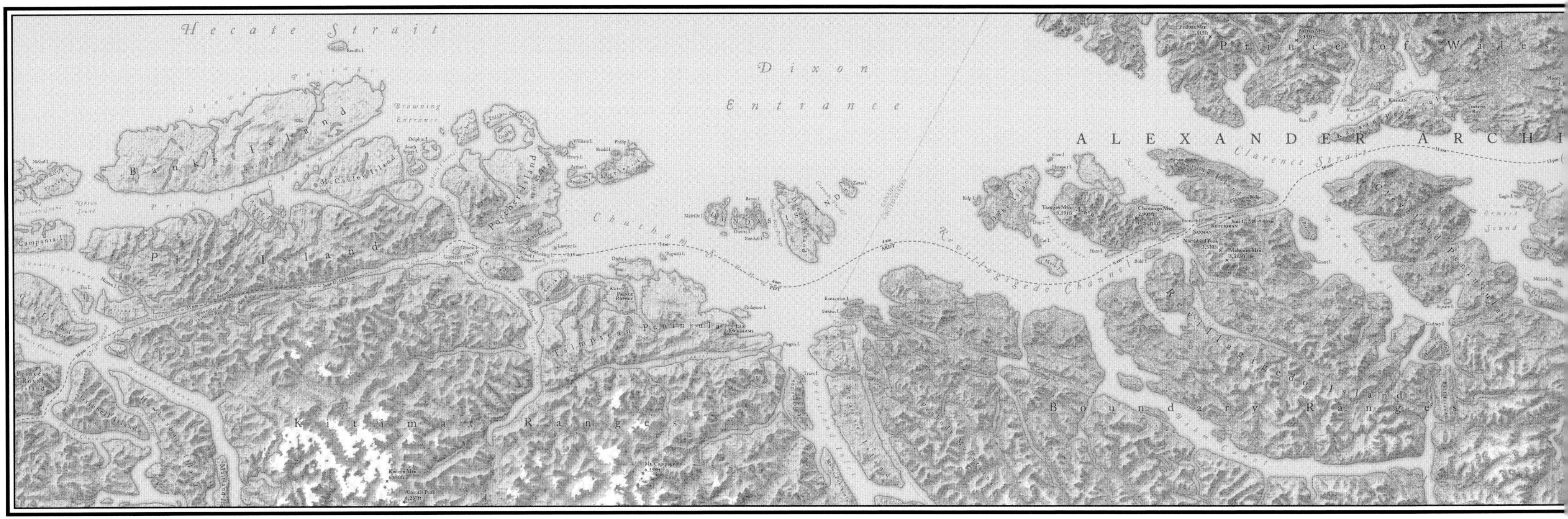
Hecate Strait
Dixon
Entrance
Banks Island
Pitt Island
Chatham Sound
Kitimat Ranges
ALEXANDER ARCHI
Clarence Strait
Revillagigedo Channel
Boundary Ranges

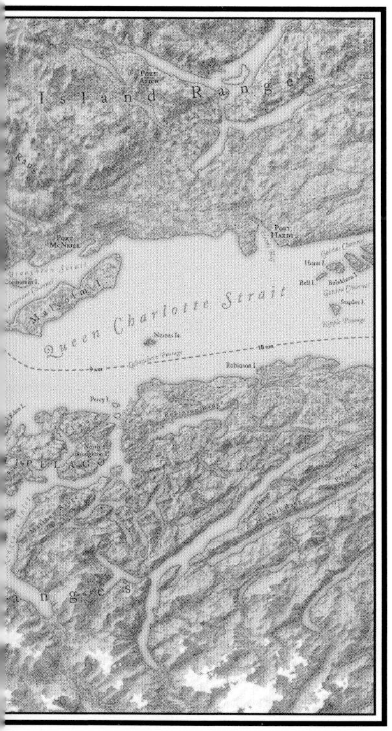

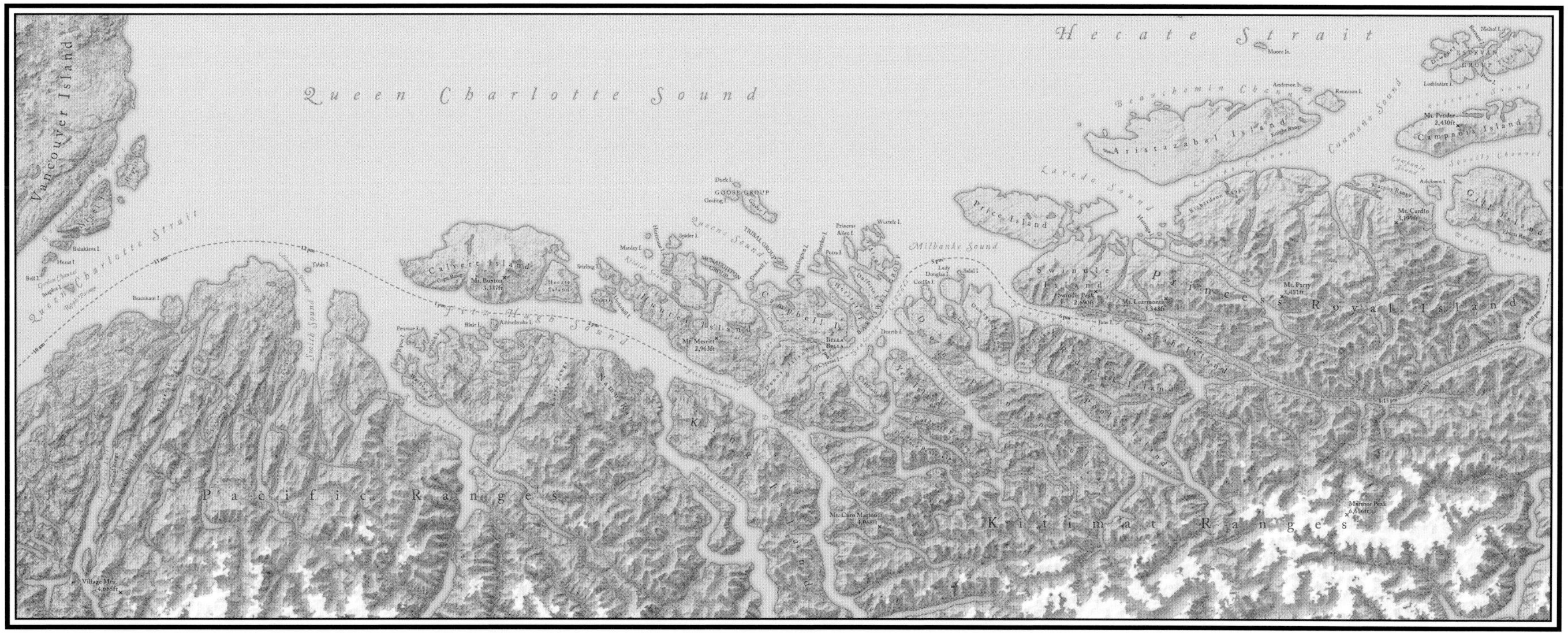

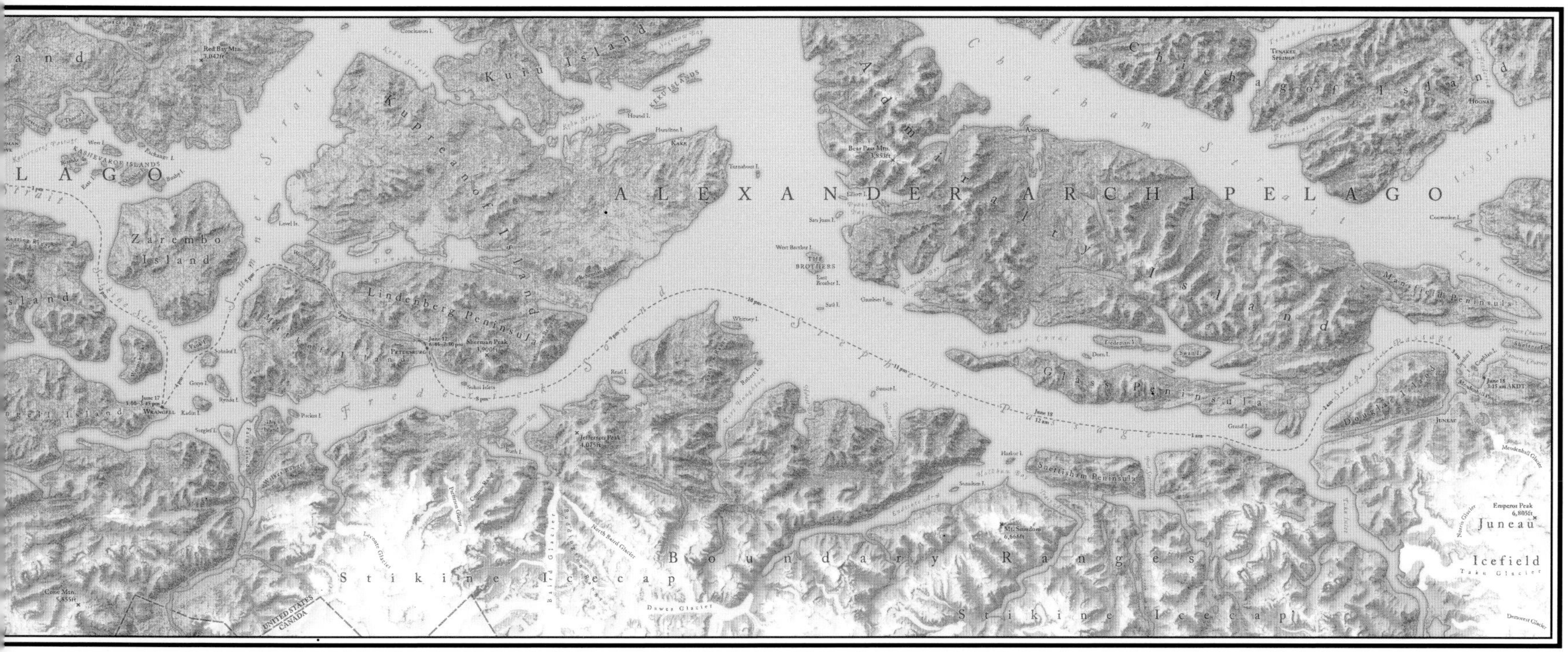

A NEW AND COMPLETE MAP OF

The Inside Passage

COVERING THE JOURNEY FROM

Bellingham to Juneau

INCLUDING THE PRINCIPAL ISLANDS,
SETTLEMENTS, PASSAGES, &c.

PREPARED IN JUNE
OF THE YEAR
TWO THOUSAND AND EIGHTEEN
BY

Daniel P. Huffman
MADISON, WISCONSIN

0 5 10 15
Scale of Miles

GREENLAND

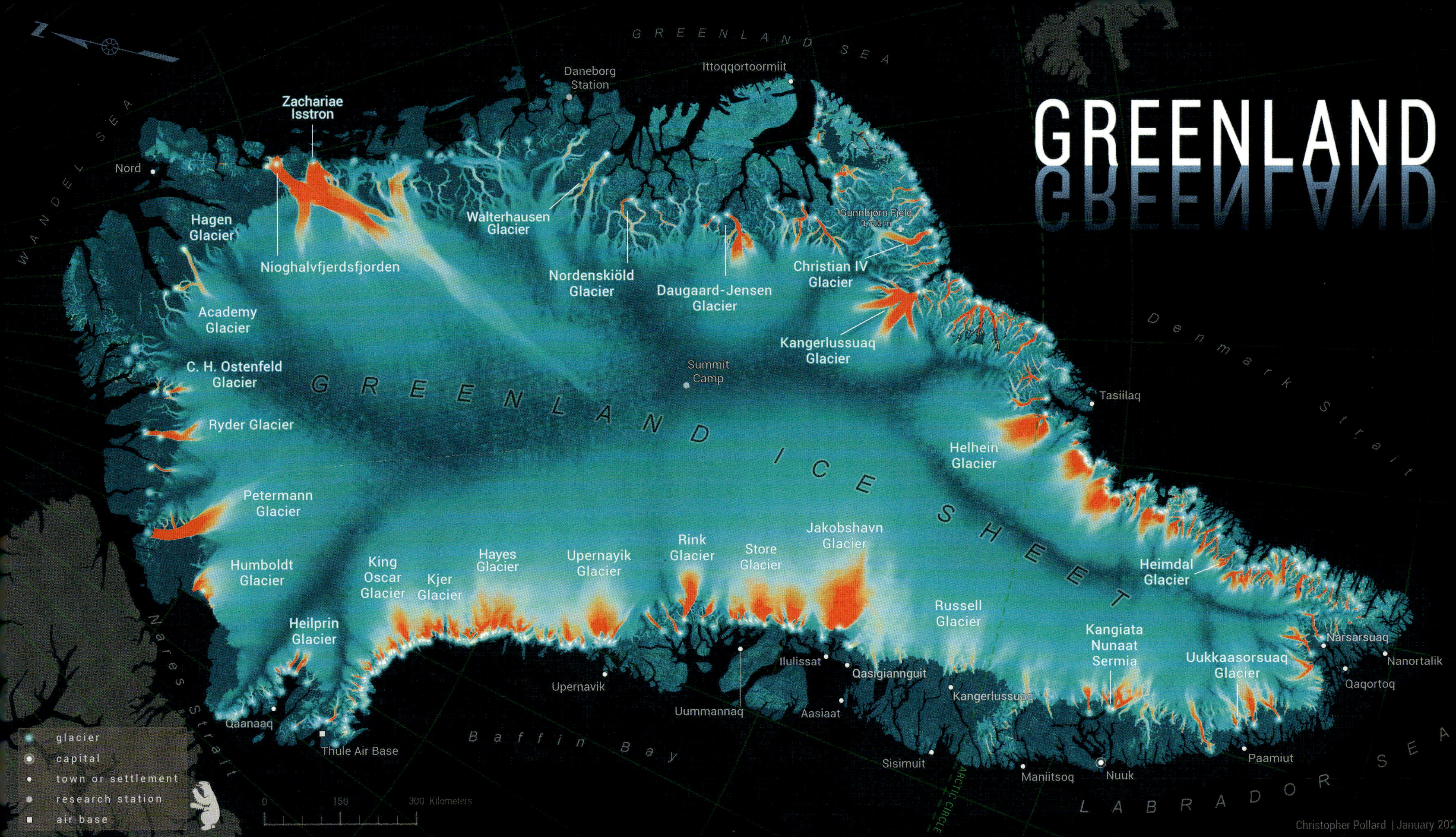

WHERE THE ICE MEETS THE OCEAN

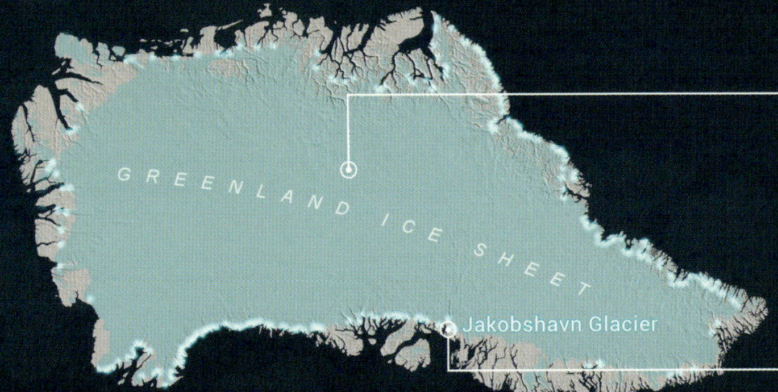

Greenland has 10 percent of the world's fresh water locked up in it's ice sheet. If fully melted, it could raise sea levels by up to 25 feet.

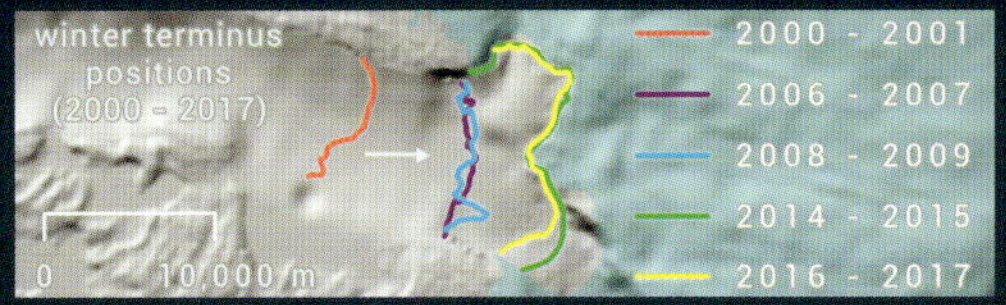

Scientists have recently discovered that Greenland's coastal glaciers have begun to melt faster over the past decade not, just due to rising air temperatures, but also because of warming ocean temperatures.

Ice Velocity Magnitude *(meters per year)*

Change in velocity should be used as an indication of relative quality

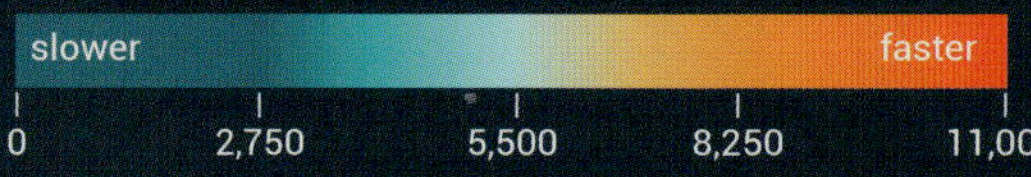

Ice velocity derived from Sentinel-1 SAR (2017-09-01 to 2018-05-31)

Sources: Joughin, I., B. Smith, I. Howat, and T. Scambos. 2015, updated 2018. MEaSUREs Greenland Ice Sheet Velocity Map from InSAR Data, Version 2. greenland_vel_mosaic200_2017_2018_vv_v02.1. Boulder, Colorado USA. NASA National Snow and Ice Data Center Distributed Active Archive Center. doi: https://doi.org/10.5067/OC7B04ZM9G6Q. 1/18/2020.

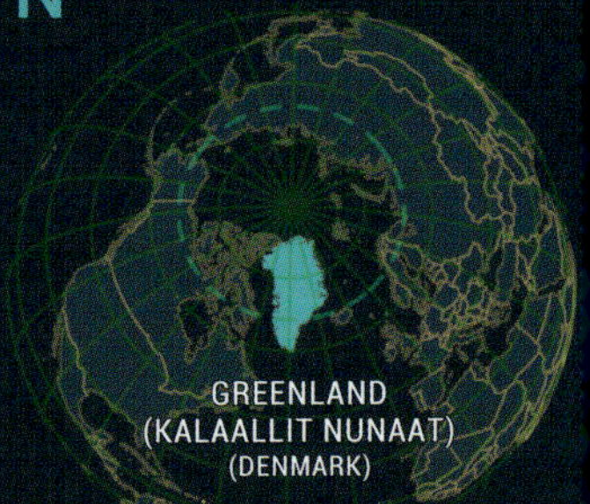

Geospatial Data | Natural Earth, NOAA, NSIDC, USG

I wanted it to feel like the title was melting into the ocean... «

« Greenland (Where the Ice Meets the Ocean)

Christopher Pollard

The origins of this map go back to November 2019, during Topi Tjukanov's (@tjukanov) #30daymapchallenge on Twitter. I joined this challenge with two objectives in mind: first, to push myself outside of my usual work habits, and second, to experiment with visualizing geospatial data in a different tone—quick, lively, colorful maps that would expand my creative palette. As my daughter so poignantly stated, "Dad is making art with his maps again."

The first iteration simply depicted the dramatic coastline, along with a few sparse towns. However, what I most enjoyed designing was the gradient-like reflection in the title. I wanted it to feel like the title was melting into the ocean, which according to scientific evidence is actually happening to the coastal glaciers on Greenland. Another feature that I wanted to highlight was the placement of Greenland in a horizontal position, as it is a visual that we don't normally see.

The map received positive reactions on Twitter, but I felt something was missing and knew I needed some compelling scientific data about ice melt to tell the visual story I had imagined.

I then explored the datasets of the National Snow and Ice Data Center (NSIDC) and knew immediately that the land ice velocity imagery was what I was seeking. I altered the map's focus to highlight the change of ice velocity occurring at the intersection of the coastal glaciers and the ocean. Initially, I began using softer shades of blue and white to represent dirty melting ice, but I wanted something more striking. With the help of my daughter, we arrived at a teal-to-orange color ramp that helped define and accent the change in velocity. This change can be seen clearly where the glaciers meet the ocean.

Finally, it was time to label the glaciers and other features. Although challenging and very time-consuming, the task of figuring out label placement proved to be the most rewarding part.

And in the end, our goal as mapmakers is to combine accurate data features with a stunning piece of artwork, to create a visual feast for the eyes.

Island Time »

Graphics department collective under the direction of Delphine Papin — Le Monde

Le Monde's graphics department created these maps for the "Island Time" exhibition at the Museum of European and Mediterranean Civilizations. The commission from the exhibition's curators was to show that each island, worldwide, is not isolated and actually takes part in a large network. We established a list of ten islands with environmental, migratory, or military issues, while emphasizing the interweaving of the geographical scales.

We also worked to create a graphic design that would capture a museum visitor's attention, as well as help them understand complex issues in a simple way. We wanted to emphasize each island without overwhelming the viewer with too much information, and we created each map at poster size. The stakes were very different from our usual design situation of a classic newspaper, as the posters had to be legible from a few yards away, with sleek black-and-white pictograms.

It took several weeks of creative design work to prepare the production of the exhibition maps. We chose two colors, a blue-green and a dark pink, reminiscent of the sea colors at different times of the day, in a Claude Monet manner. We also drew inspiration from the series of atlases produced by Arthaud Editions, whose maps are predominantly dichromatic.

The coastal ripples are another central element of our design. They represent the ripples formed by ebbing water receding from the shore in concentric waves, evoking the maps from past centuries that used this design principle to add emphasis to the oceans, otherwise quite devoid of information and visually bland.

Lastly, our signature style resides in the legend linking the two maps. Instead of a "traditional" legend, in which the different elements are merely listed, our legend integrates them into a text that reads like a narrative, in the fashion of books for children learning to read. Here, it's the visitor whom we accompany into reading the map and the world.

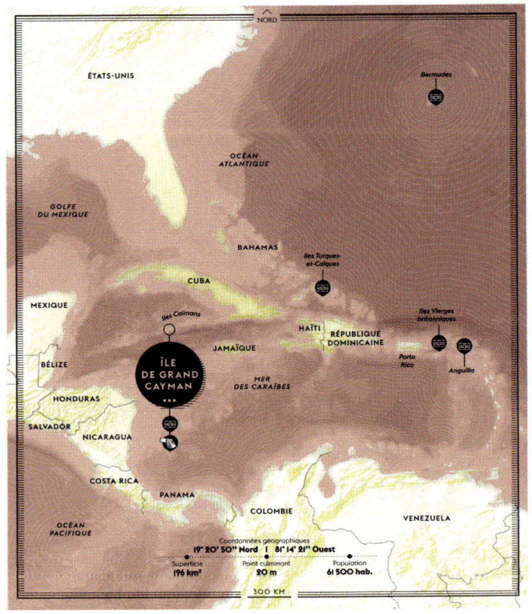

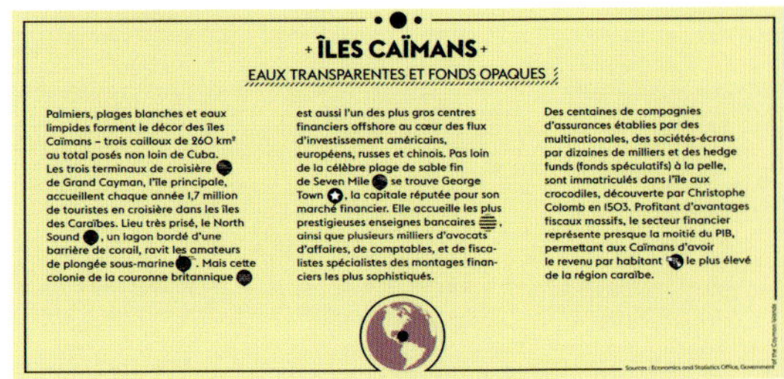

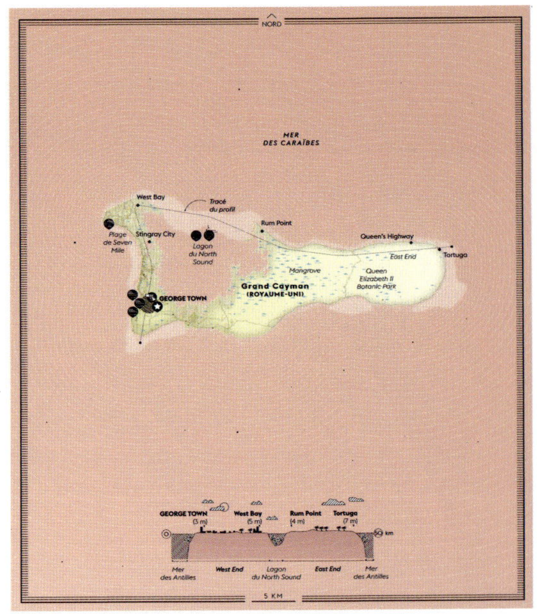

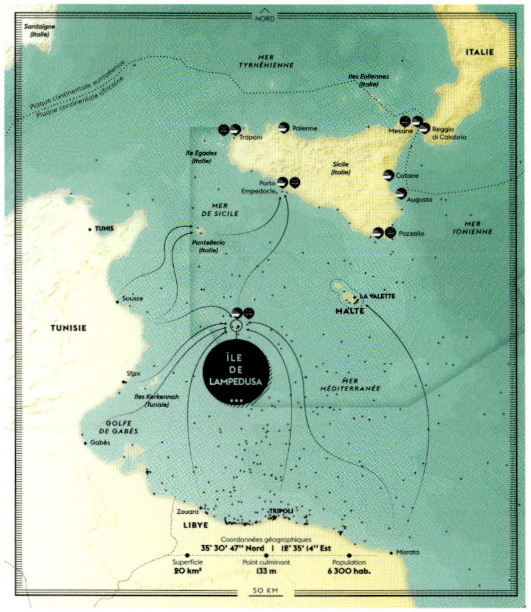

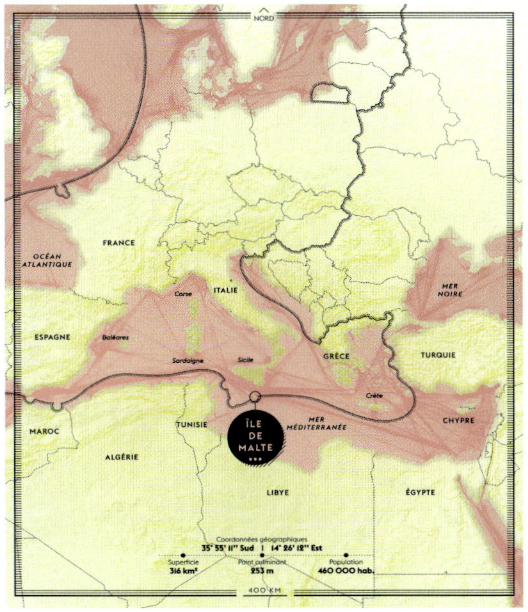

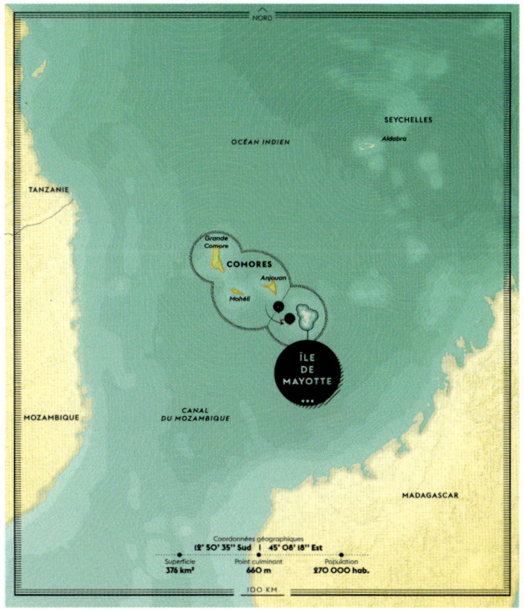

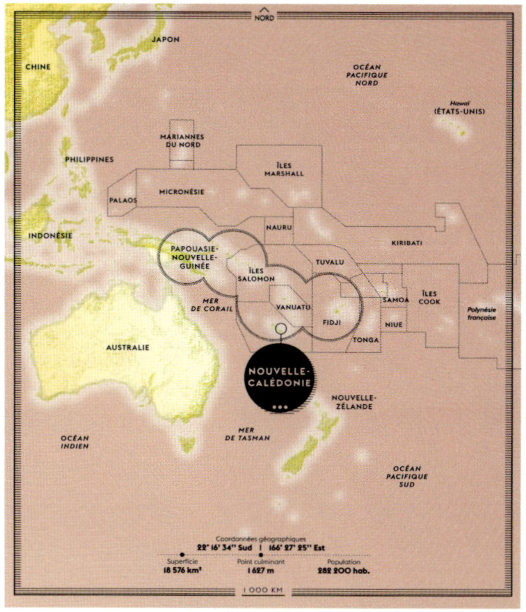

LAMPEDUSA

PORTE FERMÉE DE L'EUROPE

Située à l'extrême sud de l'Europe, plus proche des côtes africaines que de l'Italie, dont elle dépend, Lampedusa est intimement liée, depuis les « printemps arabes », à la question migratoire. Cette île vit de la pêche et du tourisme, avec ses plages de sable blanc, sa réserve naturelle et son centre de soin pour les tortues marines. Mais, depuis 2011, les contrecoups géopolitiques de la révolution tunisienne, puis ceux de la guerre en Libye, ont fait de Lampedusa l'avant-poste de la forteresse qu'est l'Union européenne (UE). Dans la nuit du 2 au 3 octobre 2013, le naufrage d'un bateau a fait 366 morts, une catastrophe d'une ampleur alors inédite. Ce drame entraîne la création de « Mare Nostrum », une opération de sauvetage mise en place par l'Italie pour intervenir en mer, sur la voie des exilés →. Elle s'arrête un an plus tard, sous la pression des partenaires européens craignant un appel d'air. L'UE renforce ses contrôles et fait de Lampedusa l'un des « centres de tri » pour migrants. Les structures d'accueil de l'île sont vite surchargées : plus de 600 000 personnes sont arrivées en Italie par la Méditerranée depuis 2014. Alors que plus de 15 000 migrants auraient perdu la vie en tentant la traversée sur cette même période, Rome ferme désormais ses ports aux bateaux de sauvetage. La « porte de l'Europe », du nom de la sculpture érigée en mémoire des migrants disparus en mer, est devenue une salle d'attente.

Sources : collettivo Askavusa ; Caritas Italiana ; Haut Commissariat des Nations unies pour les réfugiés (HCR)

MALTE

ÎLE FRONTIÈRE AUX CONFINS DU LICITE

Jadis escale pour les commerçants en route vers l'Orient, lieu de brassage des cultures méditerranéennes, Malte défraye la chronique avec ses scandales de corruption. Après le départ de l'autorité coloniale britannique, en 1964, le tout jeune Etat doit bâtir son indépendance économique. Avec un territoire exigu, sans ressources agricoles ou minières, l'archipel se tourne alors vers le tourisme et l'activité portuaire. Profitant d'une géographie exceptionnelle, au cœur de la Méditerranée et des routes commerciales, porte d'entrée de l'Europe et de l'espace Schengen, Malte a fait de son port franc une place stratégique. Son économie florissante repose aussi sur les places financières concentrées dans la capitale, qui agissent à la frontière de la légalité. Le plus petit Etat de l'Union européenne (UE), mais aussi le plus dense en population, attire aujourd'hui les riches étrangers avec son large éventail de services d'« optimisation fiscale » : pavillons de complaisance, vente de passeports contre investissements dans l'immobilier, immatriculation de sociétés et de yachts, jeux d'argent… L'île permet de mettre un pied dans l'UE, tout en bénéficiant d'avantages fiscaux. Enfin, l'assassinat, en 2017, d'une journaliste locale enquêtant sur des affaires impliquant le pouvoir a brutalement remis en lumière la place de Malte dans les réseaux de la criminalité et de l'illicite.

Sources : Nathalie Bernardie-Tahir, professeure de géographie à l'université de Limoges ; Malta Financial Service Authority

MAYOTTE

UN BOUT DE FRANCE DANS L'ARCHIPEL DES COMORES

L'archipel des Comores, qui est une ancienne colonie française, est constitué de quatre petites îles situées à l'entrée du canal du Mozambique. Aujourd'hui, trois d'entre elles forment l'Union des Comores, tandis que Mayotte a souhaité rester française – elle est devenue département en 2011. Les habitants de ces îles entretiennent des liens familiaux forts et partagent langue et religion. Mais, depuis que la France a instauré, en 1995, un visa pour les Comoriens, une immigration illégale à Mayotte ne cesse de prendre de l'ampleur. Plus de 12 000 personnes ont péri en mer en souhaitant rejoindre l'île sur des petites embarcations de pêche, les kwassa-kwassa. Depuis cinq ans, l'écart du niveau de vie s'est amplifié, et les migrations se sont intensifiées. En 2017, près d'un habitant sur deux est de nationalité étrangère, principalement comorienne, et les trois quarts des bébés nés à la maternité de Mamoudzou sont de mères étrangères. Mobilisée pour contrôler ses frontières, la France a mis en place des radars, des vedettes et des bateaux intercepteurs, et effectue de nombreuses interpellations de sans-papiers sur l'île. Ainsi, 15 500 migrants (à 98 % comoriens) ont transité en 2017 par le centre de rétention administrative de l'île.
Mais Mayotte est aussi une terre d'émigration, les Mahorais cherchant eux aussi une meilleure qualité de vie, à La Réunion et en métropole.

Sources : Migreurop ; INSEE ; La Cimade

NOUVELLE-CALÉDONIE

UNE SOCIÉTÉ DIVISÉE, À LA RECHERCHE D'UN DESTIN COMMUN

La Nouvelle-Calédonie appartient aux îles mélanésiennes, qui s'étalent en chapelet au large de l'Australie. Elle devient une colonie française de peuplement en 1853. Ses premiers habitants, les Kanak, sont cantonnés dans des réserves sous le régime de l'indigénat, aboli en 1946. Le recours à l'immigration lors du boom du nickel des années 1970 rend les Kanak minoritaires et engendre un réveil identitaire. Ils revendiquent l'indépendance de l'île et s'opposent aux loyalistes – des affrontements font plus de soixante-dix morts entre 1984 et 1988. La prise d'otages de la grotte d'Ouvéa amène les différents partis à négocier.
Les accords de Matignon (1988), puis de Nouméa (1998), prévoient un référendum sur l'indépendance et insistent sur la construction d'un destin commun avec la création d'un centre culturel kanak, une réforme foncière et un rééquilibrage économique. La société néo-calédonienne reste aujourd'hui encore très inégalitaire. Elle souffre également d'une forte ségrégation aussi bien à l'échelle de l'île – une grande part du territoire est peuplé à 70 % de Kanak – qu'à Nouméa, où vivent les deux tiers de la population. Lors du référendum de 2018, les Kanak ont massivement voté pour l'indépendance, même si le non l'a emporté à 57 %. Le nombre d'habitants ne se déclarant pas appartenir à une communauté, mais se définissant comme « calédoniens », est en augmentation. Ce sentiment serait-il le signe d'une société qui trouve son identité commune ?

Sources : Atlas de la Nouvelle-Calédonie, IRD-Congrès, 2013 ; IEOM ; Insee ; Ministère de l'Intérieur

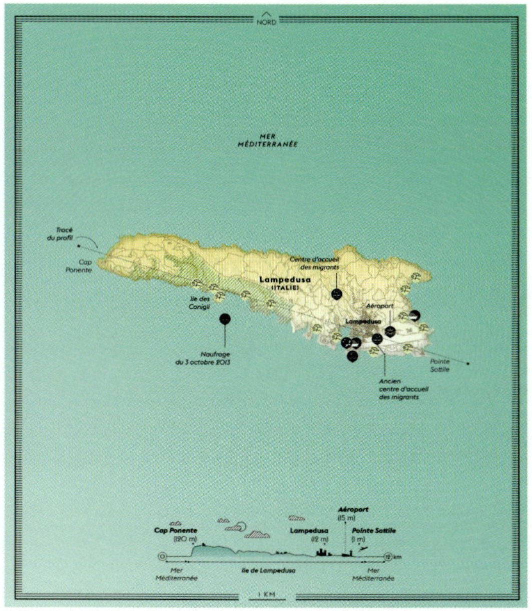

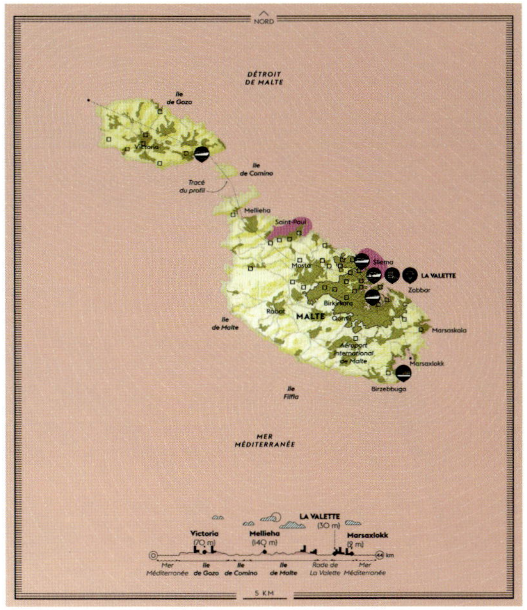

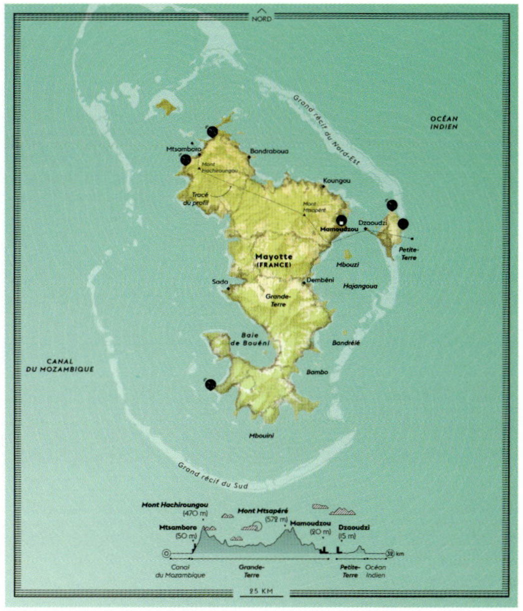

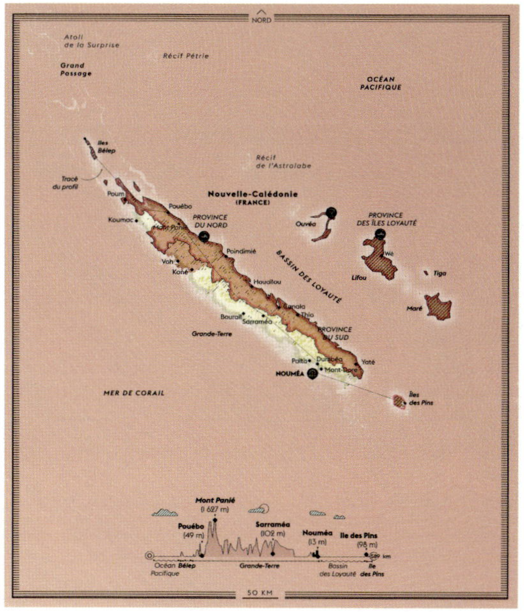

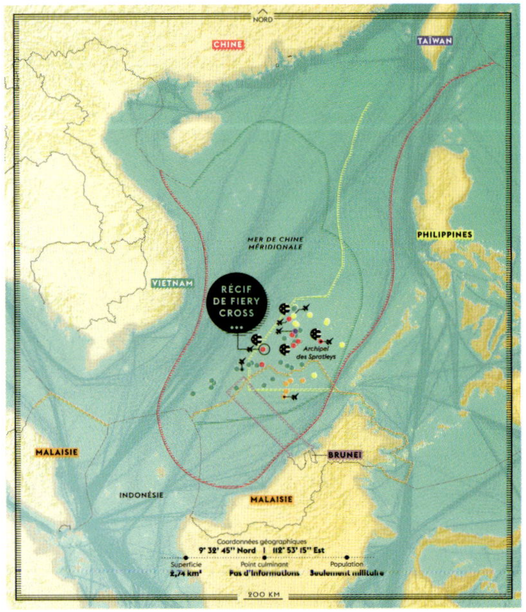

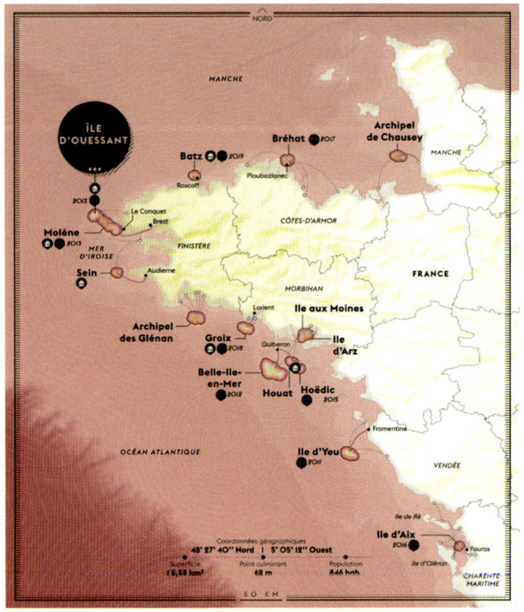

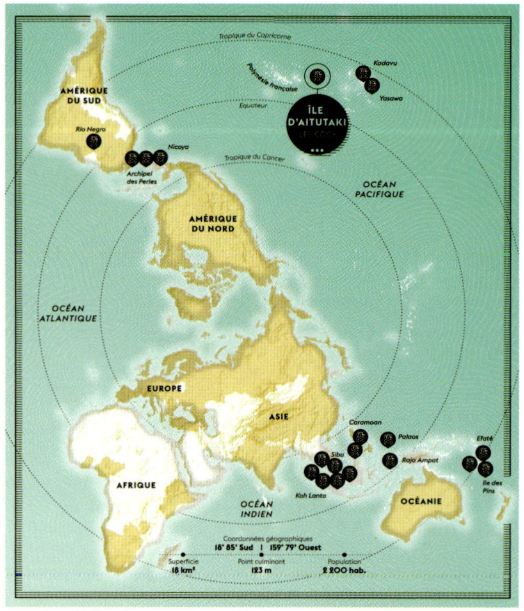

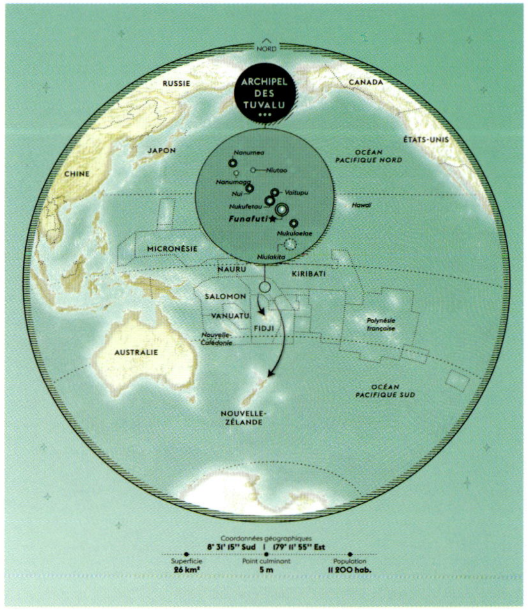

SPRATLEYS

CHÂTEAUX DE SABLE EN MER DE CHINE

Au cœur de la mer de Chine méridionale, les îlots et récifs coralliens des Spratleys, quasi inhabités, sont pourtant convoités par au moins cinq pays. La Chine s'emploie à agrandir ces îles, voire à en faire émerger de nouvelles, pour y déployer ses moyens militaires. Ses rivaux font de même, chacun cherchant à étendre sa zone économique exclusive. L'enjeu est de taille : contrôler la route commerciale maritime où transite un tiers du commerce mondial, les eaux poissonneuses et les fonds marins potentiellement riches en hydrocarbures.

Le récif de Fiery Cross est emblématique de cette compétition insulaire. En 1988, la Chine y installe une station météorologique sur une plate-forme de béton. En 2014, elle y entame une transformation spectaculaire en faisant une île artificielle de près de 3 kilomètres carrés, avec une piste d'atterrissage, un port et de multiples dispositifs militaires.

Cet expansionnisme chinois inquiète les Etats-Unis, qui, depuis 2015, font croiser leurs navires militaires près des îlots occupés. La Chine y voit une violation de sa souveraineté. Cependant, saisie par les Philippines, la Cour permanente d'arbitrage de La Haye a jugé, en 2016, que les droits historiques sur lesquels Pékin s'appuie pour revendiquer la mer de Chine méridionale n'ont aucun fondement.

ÎLES DU PONANT

RÉSILIENCE EN MER D'IROISE

Là où le soleil se couche, le littoral français se brise en une myriade d'îles dans l'Atlantique. Quinze sont habitées : les îles du Ponant. Distantes du continent de 500 mètres à 25 kilomètres, elles sont connectées à celui-ci par des liaisons maritimes quotidiennes, intensifiées à la belle saison.

Si le nombre d'îliens a été divisé par deux depuis le début du XXe siècle, la tendance est aujourd'hui à la stabilité. Certaines îles ont enrayé la chute en attirant de nouvelles populations, séduites par un cadre de vie plus solidaire et plus proche de la nature. Longtemps isolées, les îles du Ponant fonctionnent de plus en plus en réseau. En 1975, un collège est créé avec des antennes sur six îles pour que les élèves puissent poursuivre le cycle secondaire. Depuis 2011, le festival Les Insulaires ! se réunit chaque année sur l'une des îles. En 2017, le label « Savoir-faire des îles du Ponant » est lancé pour encourager les initiatives et les productions locales. A Ouessant, l'île la plus éloignée, l'agriculture a disparu au cours du XXe siècle, laissant les terres en friche. Pour y remédier, la municipalité met à disposition des terrains où ont pu s'installer un éleveur et un maraîcher bio. L'objectif est également d'atteindre 100 % d'énergie renouvelable d'ici à 2030. Une hydrolienne couvre à hauteur de 15 % les besoins énergétiques. Associées, ces îles pèsent aujourd'hui davantage auprès des collectivités, ouvrant de nouvelles perspectives.

TÉLÉ-RÉALITÉ

DES PALMIERS INTERCHANGEABLES

Jouer aux pirates, aux Robinson ou à Adam et Eve sous le feu des caméras, dans des décors insulaires aux plages immaculées, est un phénomène en vogue depuis plus de vingt ans. « Survivor », jeu insulaire par excellence et pionnier en la matière, est construit sur l'imaginaire de la robinsonnade. Proposée pour la première fois par une télévision suédoise en 1997, puis adaptée par des franchises, l'émission est diffusée dans le monde entier. Sa version française, « Koh-Lanta », « l'île au million d'yeux » en thaï, compte plus de vingt saisons, majoritairement tournées sur des îles, archipels ou péninsules de la zone intertropicale. Dans ces théâtres naturels bleu turquoise, plusieurs équipes s'affrontent ; par exemple, la version américaine a été tournée en 2006 sur l'île d'Aitutaki, aux îles Cook, où quatre groupes de concurrents séjournent sur quatre îlots inhabités. Les défis sont filmés sur deux autres îles. Lors des « conseils tribaux », autour d'un feu, les participants votent pour s'éliminer jusqu'à ce qu'il n'en reste plus qu'un. Les vaincus se rejoignent sur l'île de l'exil. Hors champ, des équipes techniques résident plusieurs semaines sur l'île principale, la seule habitée, pour faire naître le spectacle. Quel que soit le pays acheteur ou la saison, la trame du jeu est identique, le décor et les épreuves sont similaires ; seule la dramaturgie évolue selon les protagonistes.

Ces émissions sont symptomatiques d'une standardisation occidentale de l'imaginaire insulaire et d'une marchandisation des îles, dont les spécificités sont niées, afin de mettre en scène d'illusoires aventures sur des plages interchangeables.

TUVALU

ARCHIPEL EN VOIE DE DISPARITION

Au cœur de l'océan Pacifique, à mi-chemin entre l'Australie et les Etats-Unis, émerge l'archipel polynésien des Tuvalu. Six atolls coralliens aux lagons ouverts sur l'océan, deux îles récifales aux lagons à l'intérieur des terres et une petite île sans lagon composent l'un des plus petits Etats du monde. Ce décor paradisiaque aurait pu faire la réputation de Tuvalu. Mais le pays est presque parfaitement plat – en dessous de 5 mètres d'altitude –, donc extrêmement vulnérable au réchauffement climatique et à l'élévation du niveau des mers. Depuis quelques années, les marées sont plus grandes et plus fréquentes et l'érosion grignote petit à petit les côtes. Les neuf îles ont déjà vu disparaître trois mètres de front de mer et pourraient être menacées de disparition. Cette vulnérabilité est renforcée par le manque d'eau potable : l'archipel ne possède ni lac ni rivière. En raison des inondations fréquentes et de l'érosion côtière, la salinité des sols s'est considérablement accrue, ce qui rend l'eau impropre à la consommation. Pourtant, la population ne fait qu'augmenter : Funafuti, l'atoll-capitale, est passé de 2 000 à 4 500 habitants en trente ans. Mais, face au risque climatique, certains préfèrent quitter l'archipel pour les îles Fidji et la Nouvelle-Zélande. Le combat que mènent les Tuvaluans pour sauver leur pays est devenu un emblème de la lutte contre le réchauffement. Seraient-ils les premiers réfugiés climatiques ? Tuvalu, île sentinelle, serait-elle annonciatrice de notre propre futur ?

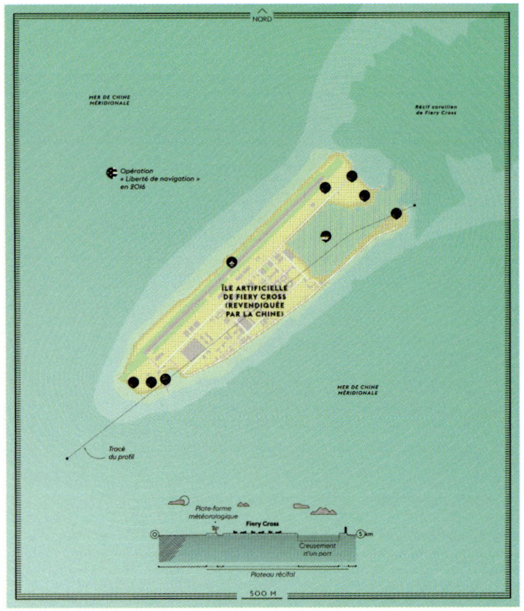

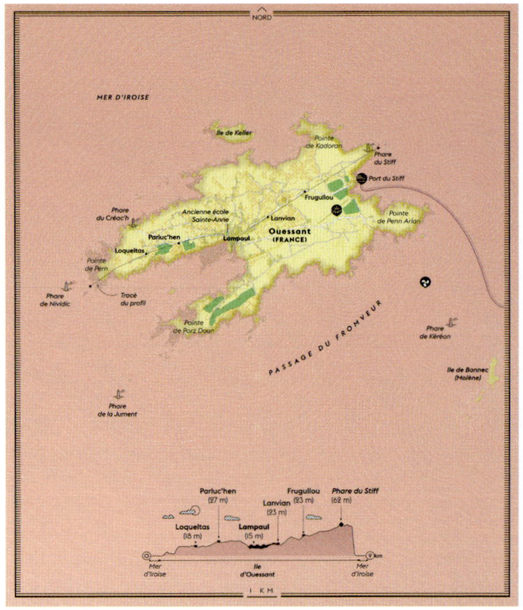

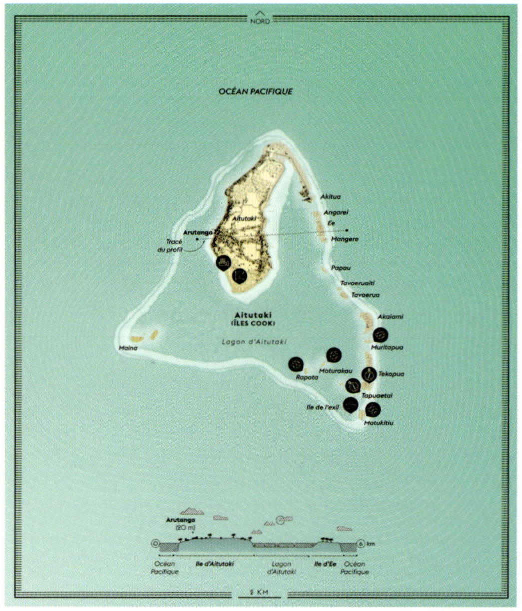

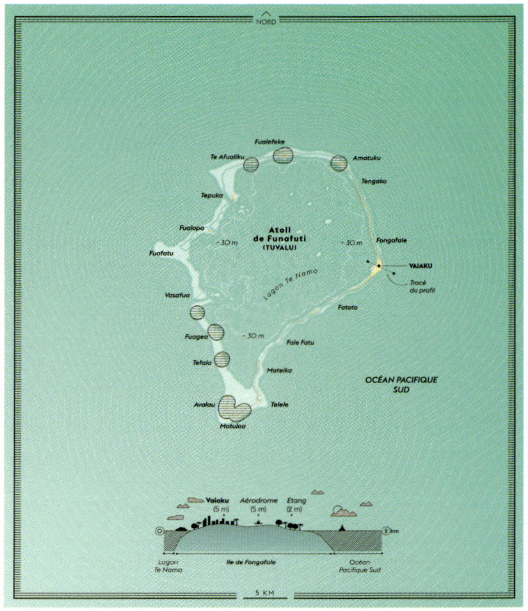

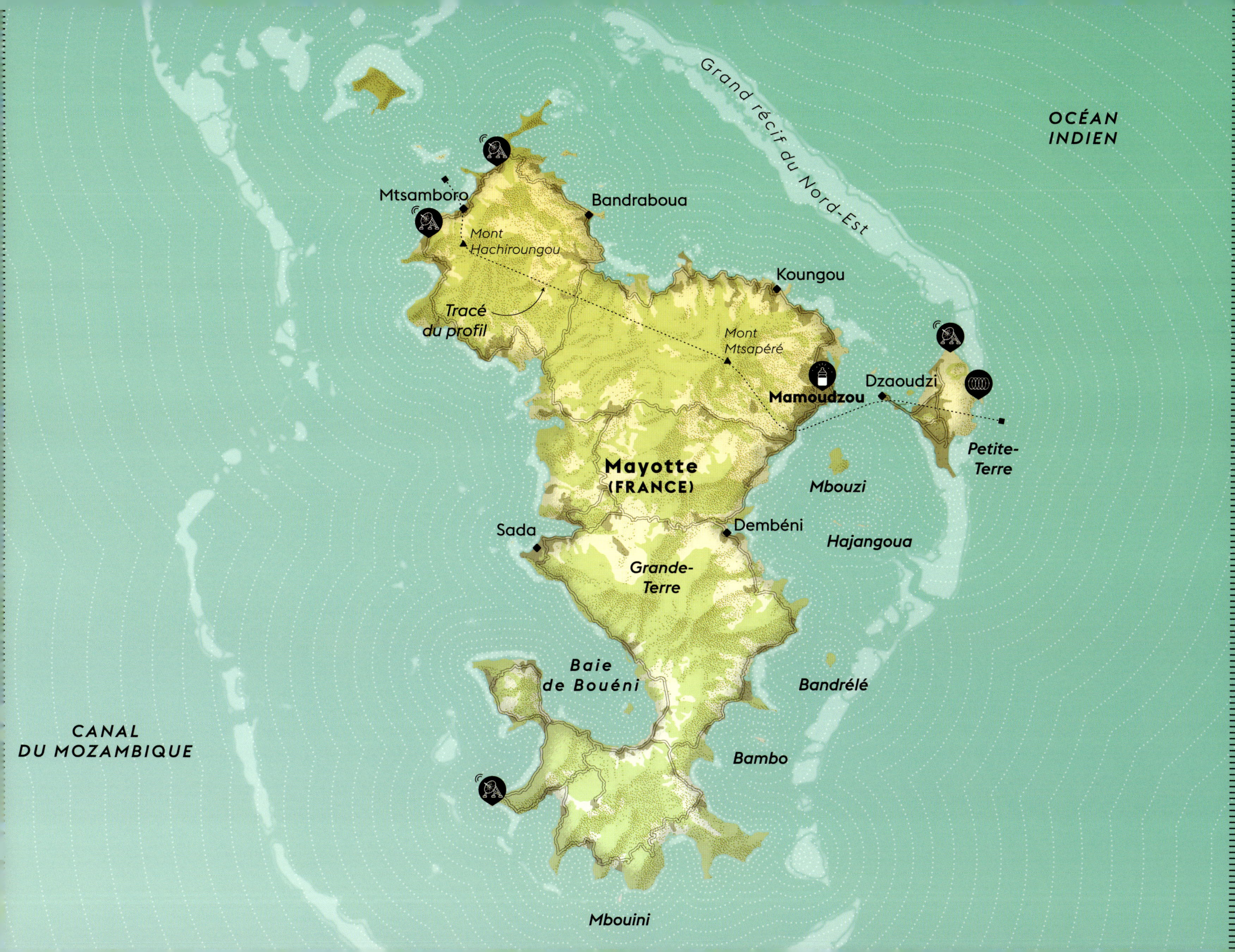
Grand récif du Nord-Est
OCÉAN
INDIEN
Mtsamboro
Bandraboua
Mont
Hachiroungou
Koungou
Tracé
du profil
Mont
Mtsapéré
Mamoudzou
Dzaoudzi
Petite-
Terre
Mayotte
(FRANCE)
Mbouzi
Sada
Dembéni
Hajangoua
Grande-
Terre
Baie
de Bouéni
Bandrélé
CANAL
DU MOZAMBIQUE
Bambo
Mbouini

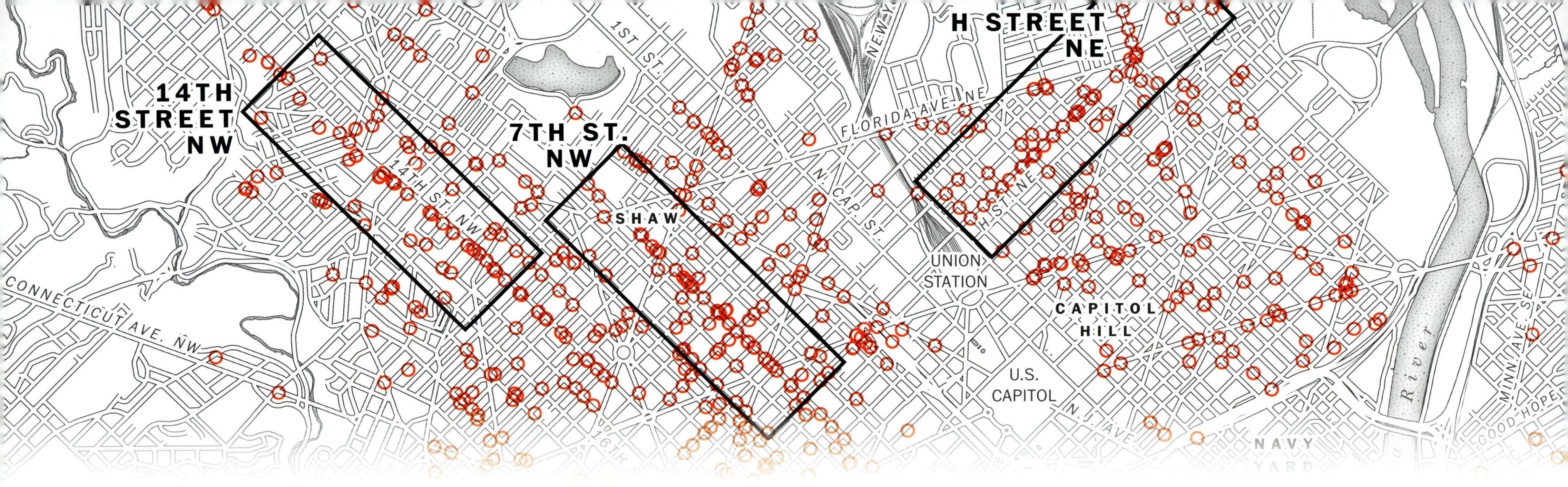

Mapping the Fiery Chaos »

Lauren Tierney, Armand Emamdjomeh, Aaron Steckelberg, Ann Gerhart, Danielle Rindler, Michael E. Ruane, and Laris Karklis — The Washington Post

Following the news of the assassination of Reverend Martin Luther King, Jr., on April 4, 1968, long-simmering racial tensions exploded throughout Washington, DC. Four days later, 13 people were dead and more than 900 businesses were damaged. Many of the mostly Black areas that were destroyed took decades to recover.

For the fiftieth anniversary of the riots, *The Washington Post* mapped the extent of the rioting throughout the city, plotting data compiled from declassified Secret Service reports and archival city-planning documents, and mixing in archival photo and video along with current interviews with people who had experienced those four days in 1968. This was part of a larger, newsroom-wide initiative, looking at 1968 as a transformative year in American history.

This map, the print version of our interactive digital piece (http://wapo.st/1968-dc-riots), was drafted and built on a three-day turnaround. We wanted to incorporate a map design that matched the 1960s "midcentury government chic" aesthetic that we saw in so many of the archival government maps we found during our research.

We incorporated design elements from these maps, such as the beautifully cased streets, the black-and-white base map with red tones for thematic data, and orientation of the map so north was tilted toward the top left corner of the page. Rotating the map in this way also allowed for a bigger and more detailed map. The final map took up the space of a full "double truck," or two-page spread, in the newspaper.

A unique little touch we added to the map was the bold, stylized north arrow, designed similar to an archival map we encountered during our research. Because we rotated the map, we wanted to be sure to make clear the orientation of the map, and this 1960s-styled north arrow seemed to do the trick.

LOGGING THE EVENTS

Shortly after 4 p.m. on April 5, the city government formally asked for federal assistance. The Secret Service began collecting reports from its field agents, and the D.C. police and fire departments. A new analysis of those records provides the most comprehensive look at the chaos intensifying and spreading through the capital.

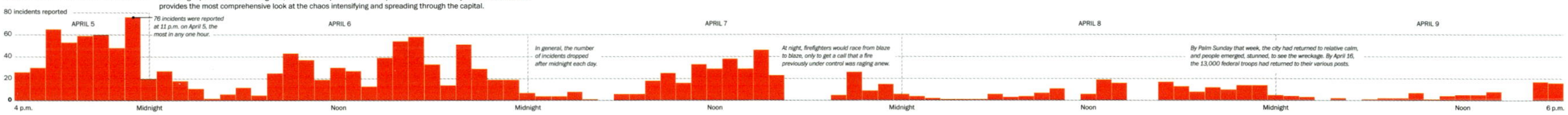

SECRET SERVICE INCIDENT REPORTS

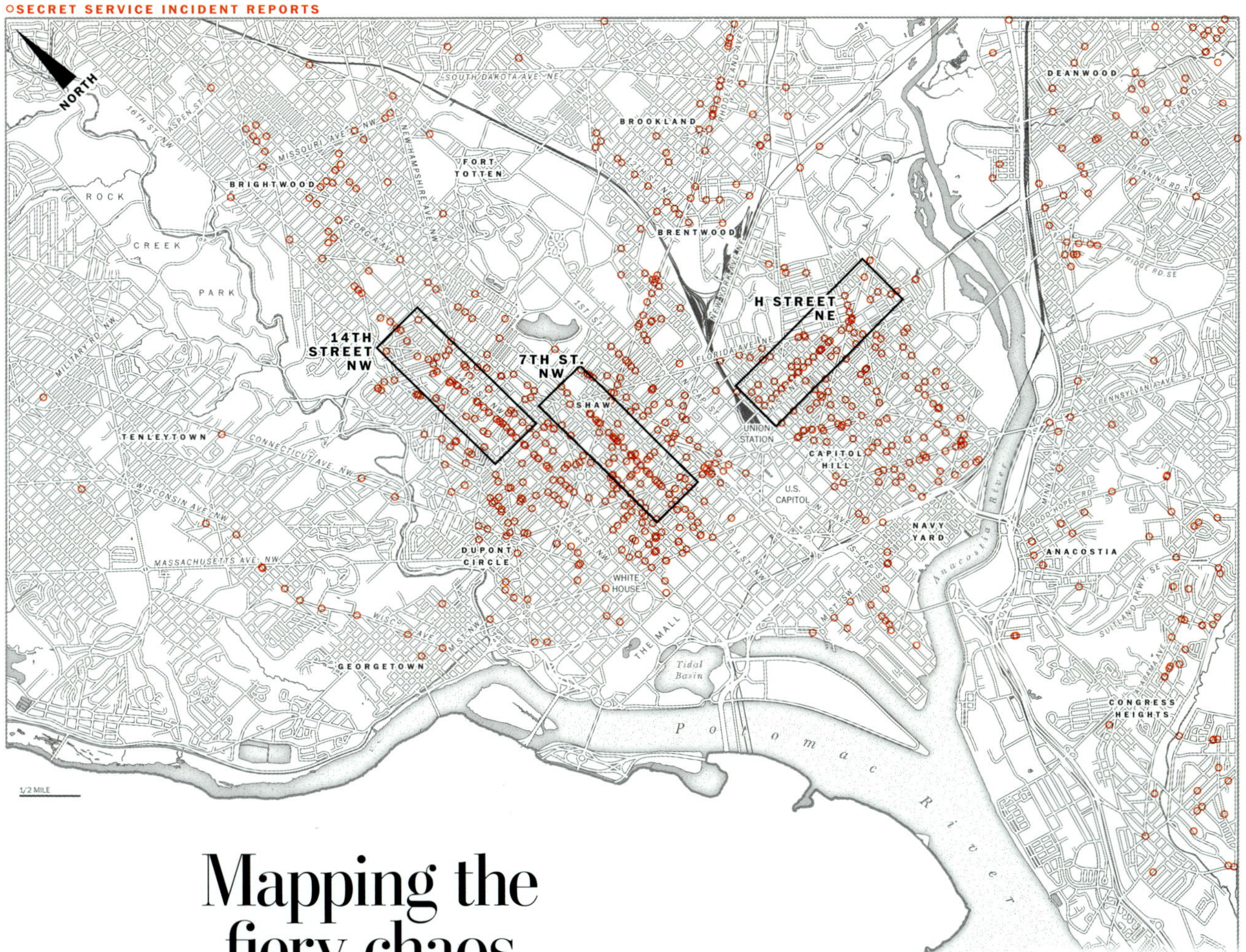

Neighborhoods reduced to rubble

The days of upheaval damaged more than 900 commercial buildings and more than 200 dwellings. An assessment by the National Capital Planning Commission noted that three shopping corridors bore the brunt of the destruction.

Buildings damaged by riots

Unsalvageable · Major damage · Minor damage

14TH STREET: THE EPICENTER OF FRUSTRATION

U Street from 14th Street east was known as the Black Broadway, a world of black theaters and jazz clubs, black lawyers and pharmacists, black newspapers and the local offices for two major civil rights groups. Some of those landmarks were spared. But the stores with nicer wares, where blacks believed they were exploited, became targets. They were stripped clean, up the hill to Columbia Heights.

SEVENTH STREET: A CORRIDOR DESTROYED

Seventh Street in Shaw was the most decimated. Half of the housing damaged throughout the city was along that corridor, and 200 of the 250 businesses along the major shopping thoroughfare were hit. Many black proprietors put "soul brother" signs on their windows, and that offered some protection. But a sign was of no help when someone fire-bombed the shop next door.

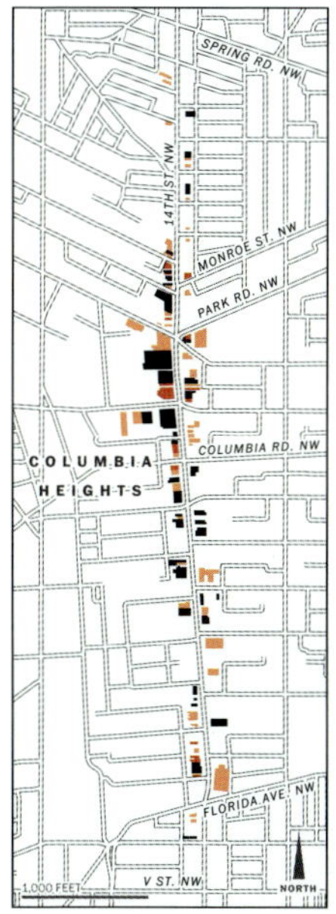

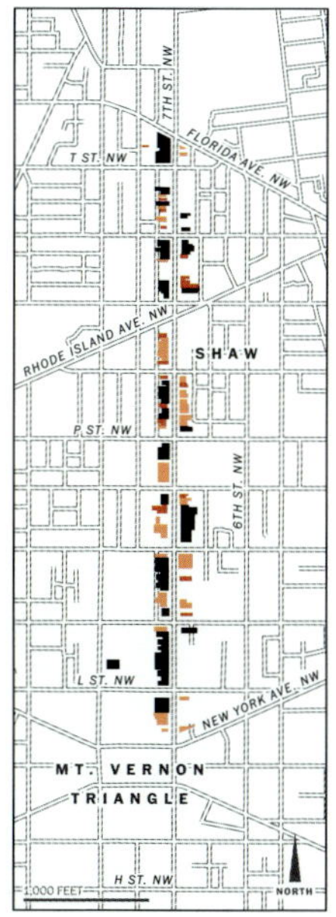

Mapping the fiery chaos

By Lauren Tierney, Aaron Steckelberg and Ann Gerhart

On April 4, 1968, the intersection of 14th and U streets in the District already was bristling with tensions between police and residents. Shortly after 8 p.m., radios broadcast the awful news: The Rev. Martin Luther King Jr. had been slain in Memphis. As night deepened, angry people began to pour from their houses into the streets, demanding businesses close out of respect.

What came next were three deadly days of rioting, burning and looting, ignited by years of frustration, anger and segregation. Large swaths of the city were reduced to rubble, for decades to come. Explore hundreds of reports in an interactive version of this map at **wapo.st/1968-dc-riots.**

H STREET: A MILE OF RUBBLE

H Street NE was a neighborhood shopping district that ran from Second to 15th streets and was the second-largest in the District. At the far end were a Sears, a Super Giant and two department stores. All of it was convenient to the working-class rowhouse neighborhoods and high-rise brick apartments that extended for blocks in both directions, to Trinidad and Capitol Hill. Much of it was unsalvageable when the smoke cleared, especially the southern side of the street between Fifth and Sixth and between Eighth and 10th streets.

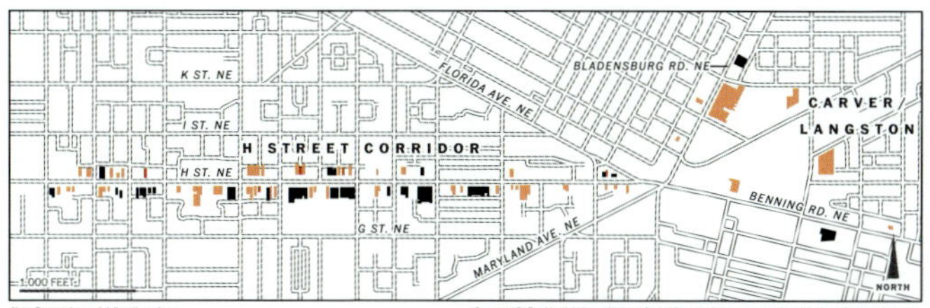

Note: Present-day road delineations shown

Sources: LBJ Presidential Library, National Capital Planning Commission and Daniel Kryder of Brandeis University

Additional reporting and graphics by DANIELLE RINDLER, MICHAEL E. RUANE, LARIS KARKLIS and ARMAND EMAMDJOMEH/THE WASHINGTON POST

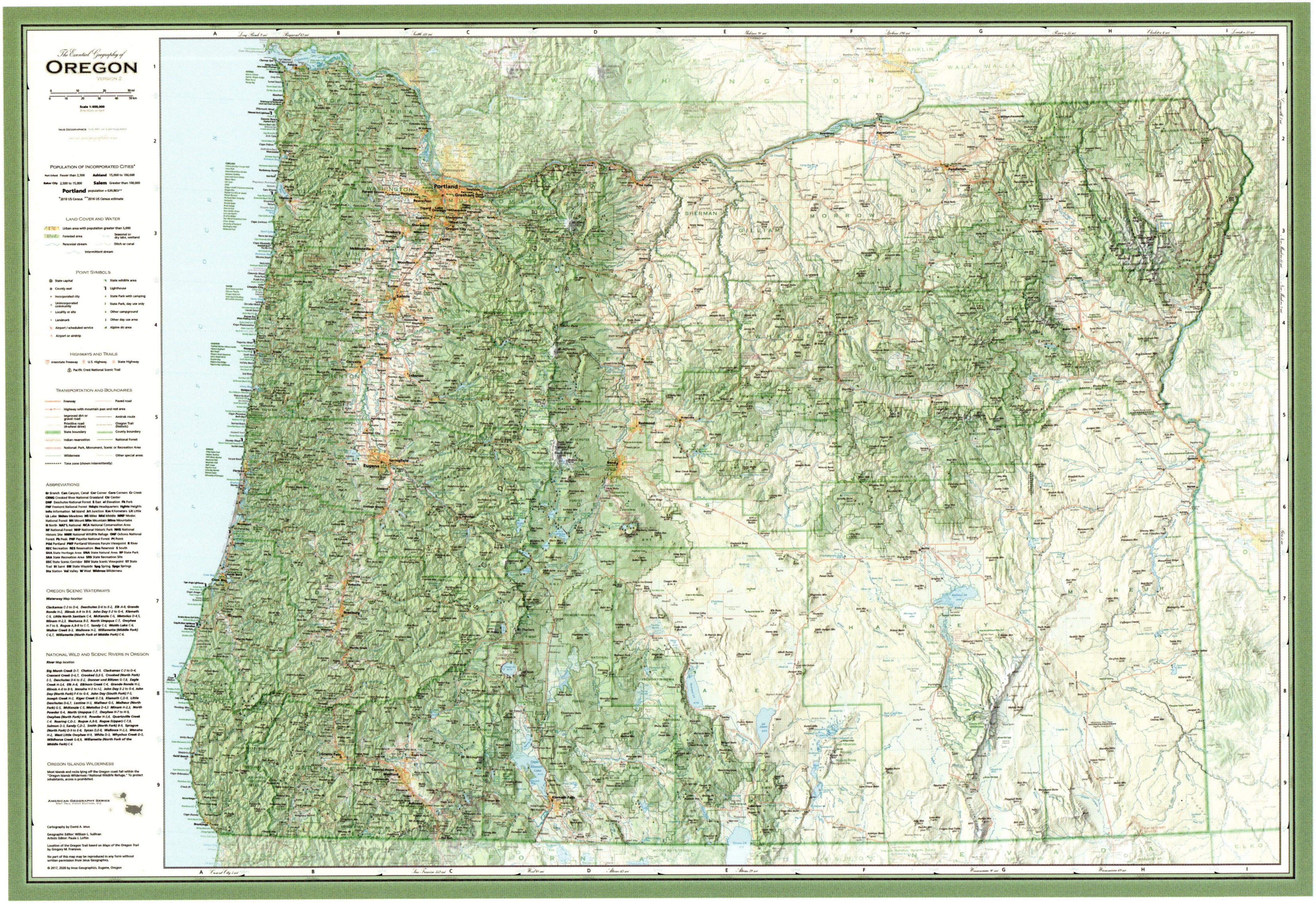
The Essential Geography of
OREGON
Version 2
Scale 1:600,000
Imus Geographics
Population of Incorporated Cities*
Port Orford Fewer than 2,500
Ashland 15,000 to 100,000
Baker City 2,500 to 15,000
Salem Greater than 100,000
Portland population = 639,863**
*2010 US Census **2016 US Census estimate
Land Cover and Water
Urban area with population greater than 5,000
Forested area
Seasonal or dry lake, wetland
Perennial stream
Ditch or canal
Intermittent stream
Point Symbols
State capital
State wildlife area
County seat
Lighthouse
Incorporated city
State Park with camping
Unincorporated community
State Park, day use only
Oregon Islands Wilderness
Most islands and rocks lying off the Oregon coast fall within the "Oregon Islands Wilderness / National Wildlife Refuge." To protect inhabitants, access is prohibited.
American Geography Series
Cartography by David A. Imus
Geographic Editor: William L. Sullivan
Artistic Editor: Paula J. Loftin
Location of the Oregon Trail based on Maps of the Oregon Trail by Gregory M. Franzwa.
No part of this map may be reproduced in any form without written permission from Imus Geographics.
© 2017, 2020 by Imus Geographics, Eugene, Oregon
Portland
Salem
Eugene

An annotated illustration rather than a map, this work drapes clear graphics over a terrain image that looks like the painted one in my vision... «

« The Essential Geography of Oregon

Dave Imus

Several years ago, I envisioned my cartography draped over terrain that had been watercolor-painted in the aesthetic of botanical art. In my vision, the terrain was as beautiful as a wildflower, and the simplicity and softness of the painted image complemented the sharpness of the graphics, creating visual harmony in which all "voices" could be "heard."

So I set out to learn how botanical artists use color and light to capture the essence of plant life. Now, using what I've learned, I create terrain illustrations by layering colored images on a computer in the way watercolorists layer paint on paper. On *The Essential Geography of Oregon*, an undershade defines broad landforms, and duotone-shaded relief illuminates local details. Nine forms of enhancement, ranging from purple mountaintops to blue-tinted lowlands, add depth, realism and a human touch, three qualities that bring geography to life.

To draft lines and place type, I use a computer almost entirely as a power-assisted extension of my hand. Following Eduard Imhof's guidelines, I render each detail and place each type label with an eye toward graphic clarity. An annotated illustration rather than a map, this work drapes clear graphics over a terrain image that looks like the painted one in my vision, making the intricate interplay of land, water, and humanity easy to see, and Oregon's geographic essence easy to understand.

Other than government offices in Western Europe that create Swiss-style topography, and a handful of independent operators, the field of cartography has not yet embraced the trained artistry that characterizes disciplines such as medical and botanical illustration. As a result, in most of the world, including the entire Western Hemisphere, there is no serviceable information on geography's primary subject matter—the basic patterns on the land. I believe the lack of basic geographical information is the reason literacy in geography is uncommon. Geographical illustrations like *The Essential Geography of Oregon* offer a practical solution to the problem.

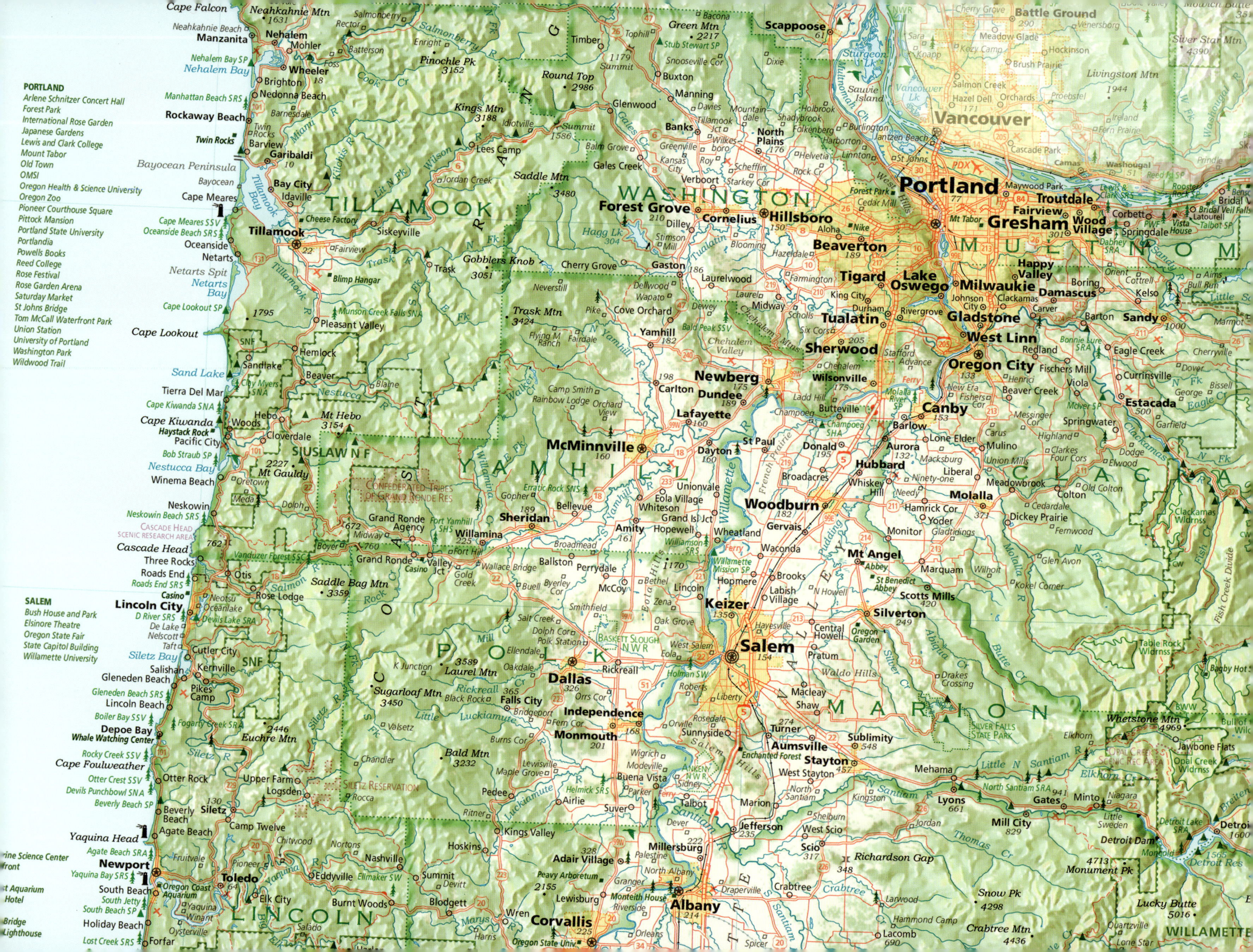

PORTLAND
Arlene Schnitzer Concert Hall
Forest Park
International Rose Garden
Japanese Gardens
Lewis and Clark College
Mount Tabor
Old Town
OMSI
Oregon Health & Science University
Oregon Zoo
Pioneer Courthouse Square
Pittock Mansion
Portland State University
Portlandia
Powells Books
Reed College
Rose Festival
Rose Garden Arena
Saturday Market
St Johns Bridge
Tom McCall Waterfront Park
Union Station
University of Portland
Washington Park
Wildwood Trail
SALEM
Bush House and Park
Elsinore Theatre
Oregon State Fair
State Capitol Building
Willamette University
Cape Falcon
Neahkahnie Mtn
Manzanita
Nehalem
Mohler
Nehalem Bay SP
Nehalem Bay
Wheeler
Brighton
Nedonna Beach
Manhattan Beach SRS
Rockaway Beach
Twin Rocks
Barview
Garibaldi
Bayocean Peninsula
Bay City
Idaville
Cape Meares
Tillamook Bay
TILLAMOOK
Cheese Factory
Cape Meares SSV
Oceanside Beach SRS
Oceanside
Netarts
Tillamook
Netarts Spit
Netarts Bay
Blimp Hangar
Cape Lookout SP
Cape Lookout
Pleasant Valley
Hemlock
Sand Lake
Sandlake
Beaver
Tierra Del Mar
Cape Kiwanda SNA
Cape Kiwanda
Haystack Rock
Pacific City
Woods
Hebo
Mt Hebo
Cloverdale
Bob Straub SP
Nestucca Bay
Winema Beach
SIUSLAW N F
Mt Gauldy
Neskowin
Neskowin Beach SRS
CASCADE HEAD SCENIC RESEARCH AREA
Cascade Head
Three Rocks
Roads End
Roads End SRS
Otis
Lincoln City
D River SRS
De Lake
Nelscott
Taft
Siletz Bay
Salishan
Cutler City
Kernville
Gleneden Beach
Gleneden Beach SRS
Lincoln Beach
Boiler Bay SSV
Depoe Bay
Whale Watching Center
Rocky Creek SSV
Cape Foulweather
Otter Crest SSV
Devils Punchbowl SNA
Otter Rock
Beverly Beach SP
Beverly Beach
Agate Beach
Yaquina Head
Agate Beach SRA
Newport
Yaquina Bay SRS
South Beach
South Jetty
South Beach SP
Holiday Beach
Lost Creek SRS
Forfar
Toledo
Siletz
Logsden
Upper Farm
Eddyville
Elk City
Burnt Woods
SILETZ RESERVATION
Euchre Mtn
Saddle Bag Mtn
Rose Lodge
Grand Ronde
Grand Ronde Agency
CONFEDERATED TRIBES OF GRAND RONDE RES
Valley Jct
Willamina
Sheridan
Bellevue
McMinnville
YAMHILL
Amity
Dayton
Lafayette
Dundee
Newberg
Carlton
Yamhill
Gaston
Forest Grove
Cornelius
Hillsboro
WASHINGTON
Banks
North Plains
Aloha
Beaverton
Tigard
Lake Oswego
Tualatin
Sherwood
Wilsonville
Milwaukie
Gladstone
West Linn
Oregon City
Canby
Portland
Gresham
Troutdale
Fairview
Wood Village
Happy Valley
Damascus
Sandy
Estacada
Molalla
Hubbard
Aurora
Woodburn
Gervais
Mt Angel
Silverton
Keizer
Salem
Turner
Aumsville
Stayton
Sublimity
MARION
Independence
Monmouth
Dallas
Falls City
POLK
Adair Village
Corvallis
Albany
Jefferson
Scio
Lyons
Mill City
Gates
Detroit Dam
Vancouver
Battle Ground
Scappoose
Sauvie Island
Kings Mtn
Saddle Mtn
Trask Mtn
Laurel Mtn
Sugarloaf Mtn
Bald Mtn
Marys Pk
Snow Pk
Crabtree Mtn
Lucky Butte
Richardson Gap
Whetstone Mtn
WILLAMETTE

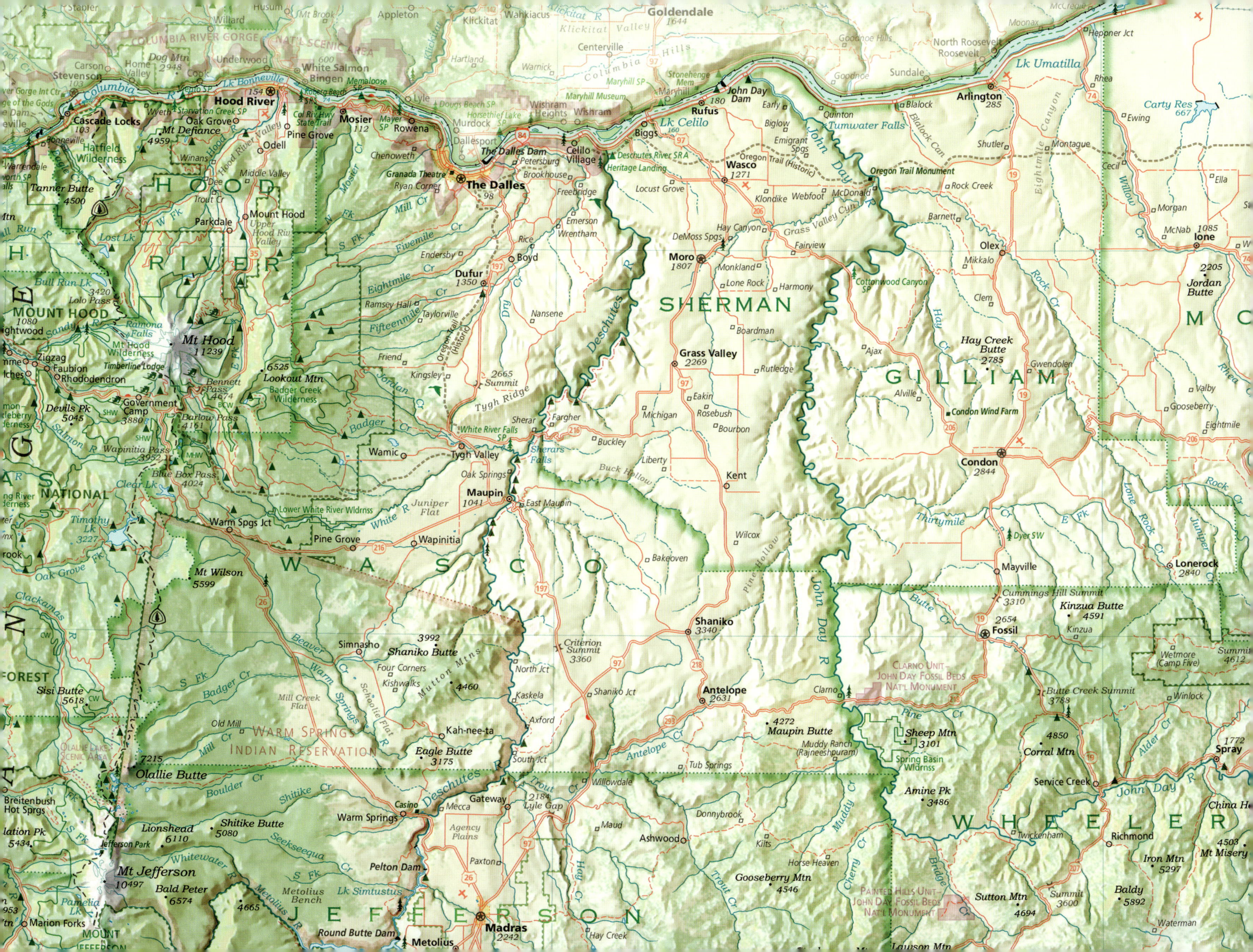

Goldendale
1644
Klickitat
Wahkiacus
Appleton
Klickitat Valley
Centerville
Columbia Hills
Hartland
Warwick
Maryhill SP
Maryhill Museum
Stonehenge Mem
Maryhill
Goodnoe Hills
Goodnoe
North Roosevelt
Roosevelt
Sundale
Lk Umatilla
Heppner Jct
Rhea
Arlington
285
Carty Res
667
Ewing
COLUMBIA RIVER GORGE
NAT'L SCENIC AREA
Willard
Husum
Mt Brook
Dog Mtn
2948
Underwood
Home Valley
Carson
White Salmon
Bingen
Stevenson
Lk Bonneville
Columbia R
Cascade Locks
Hood River
Mosier
Lyle
Dougs Beach SP
Horsethief Lake
Wishram Heights
Wishram
Murdock
Rowena
Mayer SP
Mt Defiance
4959
Oak Grove
Pine Grove
Odell
Wyeth
Starvation Creek SP
Hatfield Wilderness
Winans
Middle Valley
Dallesport
The Dalles Dam
Chenoweth
Granada Theatre
The Dalles
98
Ryan Corner
Celilo Village
Brookhouse
Petersburg
Deschutes River SRA
Heritage Landing
Freebridge
Biggs
Rufus
John Day Dam
Lk Celilo
Early
Biglow
Emigrant Spgs
Wasco
1271
Oregon Trail (historic)
Tumwater Falls
Quinton
Blalock
Blalock Can
Shutler
Montague
Cecil
Oregon Trail Monument
Rock Creek
Ella
Morgan
McNab
Ione
1085
Tanner Butte
4500
HOOD
RIVER
Mount Hood
Parkdale
Upper Hood Riv Valley
Trout Cr
Lost Lk
Bull Run Lk
3420
Lolo Pass
MOUNT HOOD
Ramona Falls
Mt Hood
11239
Mt Hood Wilderness
Timberline Lodge
Zigzag
Faubion
Rhododendron
Government Camp
Bennett Pass
4674
Barlow Pass
4161
6525
Lookout Mtn
Badger Creek Wilderness
Devils Pk
5045
Wapinitia Pass
3952
Blue Box Pass
4024
Clear Lk
NATIONAL
Timothy Lk
3227
Warm Spgs Jct
Lower White River Wldrnss
Pine Grove
Wapinitia
Emerson
Wrentham
Rice
Boyd
Endersby
Fivemile
Dufur
1350
Eightmile
Ramsey Hall
Fifteenmile
Taylorville
Nansene
Friend
Kingsley
2665
Summit
Tygh Ridge
Shearer
Fargher
White River Falls SP
Sherars Falls
Tygh Valley
Wamic
Oak Springs
Maupin
1041
Juniper Flat
East Maupin
Locust Grove
Klondike
Webfoot
McDonald
Hay Canyon
DeMoss Spgs
Grass Valley Cyn
Fairview
Moro
1807
Monkland
Lone Rock
Harmony
Cottonwood Canyon SP
SHERMAN
Boardman
Grass Valley
2269
Rutledge
Eakin
Rosebush
Bourbon
Michigan
Buckley
Liberty
Buck Hollow
Kent
Wilcox
Bakeoven
Barnett
Olex
Mikkalo
Clem
Hay Creek Butte
2785
Ajax
GILLIAM
Alville
Condon Wind Farm
Gwendolen
Valby
Gooseberry
Eightmile
Condon
2844
Thirtymile
Dyer SW
Mayville
Lonerock
2840
Jordan Butte
2205
WASCO
Mt Wilson
5599
Simnasho
3992
Shaniko Butte
Four Corners
Kishwalks
Mutton Mtns
4460
Mill Creek Flat
Old Mill
WARM SPRINGS
INDIAN RESERVATION
Kah-nee-ta
Eagle Butte
3175
Olallie Butte
7215
Sisi Butte
5618
Breitenbush Hot Sprgs
North Jct
Kaskela
Axford
Criterion Summit
3360
Shaniko Jct
Shaniko
3340
Antelope
2631
South Jct
Willowdale
Clarno
4272
Maupin Butte
Muddy Ranch (Rajneeshpuram)
Tub Springs
Cummings Hill Summit
3310
Fossil
2654
Kinzua Butte
4591
Kinzua
CLARNO UNIT
JOHN DAY FOSSIL BEDS NAT'L MONUMENT
Butte Creek Summit
3788
Winlock
Wetmore (Camp Five)
Summit
4612
Sheep Mtn
3101
Spring Basin Wldrnss
4850
Corral Mtn
Spray
1772
Service Creek
Amine Pk
3486
WHEELER
Twickenham
Richmond
China H
4503
Mt Misery
Iron Mtn
5297
Baldy
5892
Summit
3600
Sutton Mtn
4694
PAINTED HILLS UNIT
JOHN DAY FOSSIL BEDS NAT'L MONUMENT
Waterman
Lawson Mtn
Gateway
Mecca
Casino
Warm Springs
2184
Lyle Gap
Agency Plains
Paxton
Maud
Donnybrook
Ashwood
Kilts
Horse Heaven
Gooseberry Mtn
4546
Lionshead
6110
Shitike Butte
5080
Jefferson Park
Mt Jefferson
10497
Bald Peter
6574
4665
Metolius Bench
Lk Simtustus
Pelton Dam
JEFFERSON
Madras
2242
Metolius
Round Butte Dam
Marion Forks
Hay Creek
Deschutes R
John Day R

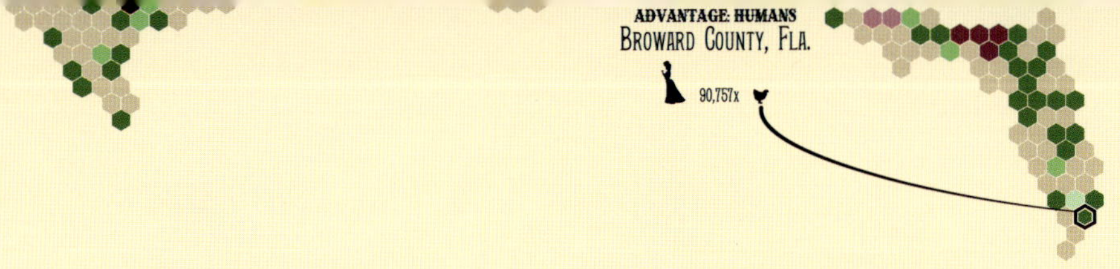

This totally normal, entirely standard-issue choropleth map with a diverging color scheme will probably only be of use in the admittedly unlikely event of nationwide conflict between humans and America's top poultry product. You know, one of those maps. But considering the fact that broiler chickens outnumber us by more than 25-to-1, should we not prepare for any scenario? Green highlights counties in which humans outnumber broiler chickens, while red represents areas where broiler chickens have the edge. Here the tile grid's uniformity negates the disproportionate prominence of extremely large counties, but does not convey any information about the absolute sizes of the human or chicken populations.

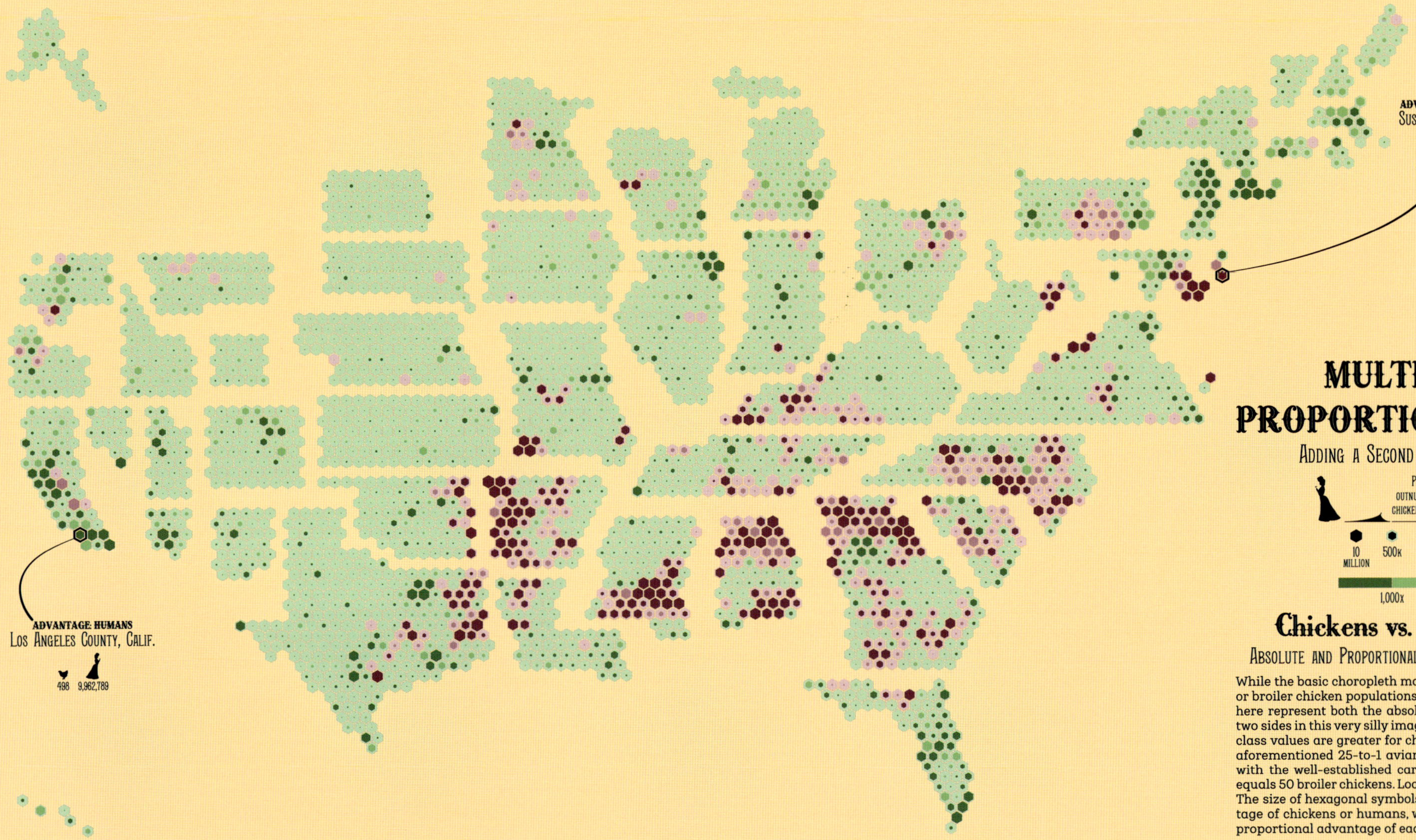

MULTIVARIATE PROPORTIONAL SYMBOLS

Adding a Second Variable to a Silly Map

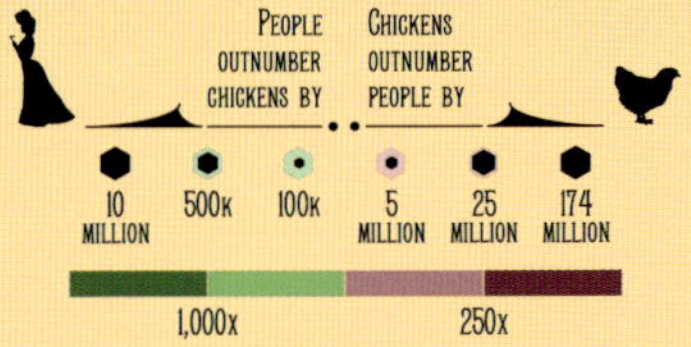

Chickens vs. Humans in 2012

Absolute and Proportional Advantage of Chickens or Humans

While the basic choropleth map above does not show the total human or broiler chicken populations, the multivariate proportional symbols here represent both the absolute and proportional advantage of the two sides in this very silly imagined conflict. Note that symbol size and class values are greater for chickens than humans to account for the aforementioned 25-to-1 avian advantage. This is entirely in keeping with the well-established cartographic convention that one human equals 50 broiler chickens. Look it up in any reputable design textbook. The size of hexagonal symbols here represents the numerical advantage of chickens or humans, while the hue and brightness reflect the proportional advantage of each species.

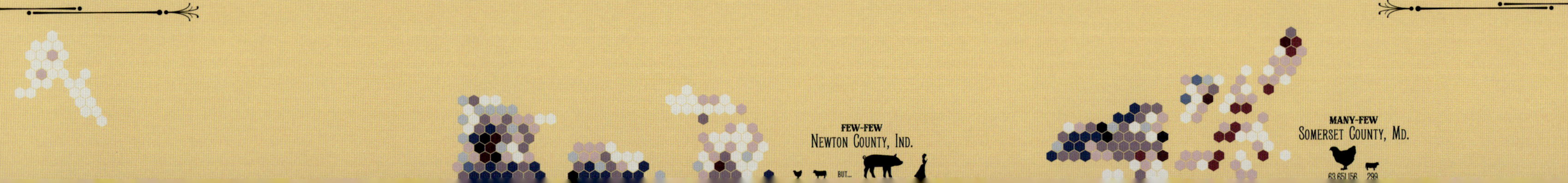

...the right amount of "weird" can actually improve the visual communication and interpretation of spatial data... «

« A Hexagonal Tile Grid of American Counties

William M. Hartnett

If there's one word not typically associated with maps produced by government agencies, "weird" must surely be near the top of the list. Sober or utilitarian, sure. Practical or useful, yes, one would hope. But the right amount of "weird" can actually improve the visual communication and interpretation of spatial data when compared to more conventional alternatives.

California's 20,000-square-mile San Bernardino County is larger than nine American states, while Hawaii's Kalawao County occupies just 12 square miles of land. The American county system is a hot mess of an administrative division, a sloppy ingredient out of which we nonetheless conjure an endless buffet of maps that don't quite measure up.

I give American counties two out of five carto-stars.

This map represents all counties and county equivalents with an identical hexagon arranged on a grid in an approximation of their real-world locations, while also preserving the relative positions and distinct shapes of most states. (Sorry, Colorado and Wyoming.) This layout is particularly well suited to the representation of qualitative data, but, as shown here, it also accommodates quantitative data in a variety of familiar visualization formats.

To introduce the hexagonal tile grid to my coworkers at the Agency for Toxic Substances and Disease Registry and Centers for Disease Control and Prevention, I used county-level statistics on broiler chicken production. In addition to being a less distracting alternative to real-world public health data, this enabled the symbological use of chicken silhouettes unfettered by basic principles of either good design or, some might say, good taste.

For viewers who might find this new or unfamiliar take on a standard map jarring, my design choices veered toward the old and reassuring to compensate. I sought to borrow from the dignified dependability that we now ascribe to the decorative flourishes and typographic extravagances of nineteenth- and twentieth-century cartography. Specifically, I imitated the ornate detail and exceptional filigree of classic Sanborn Fire Insurance Maps, hoping to convince my colleagues that this was not, in fact, weirdness for its own sake, but instead a superior interpretation of spatial information, the inevitable outcome of the irrepressible march of cartographic progress.

Even if it looks a little weird.

A HEXAGONAL TILE GRID OF AMERICAN COUNTIES

GRASP

A One-to-One Representation of Counties and County Equivalents Herein Illustrating a Variety of Visualization Methods

as made by William M. Hartnett

utilizing statistical information compiled by the

U.S. DEPT. OF AGRICULTURE
& U.S. CENSUS BUREAU

under the auspices of the

2012 Census of Agriculture
& Population Estimates Program

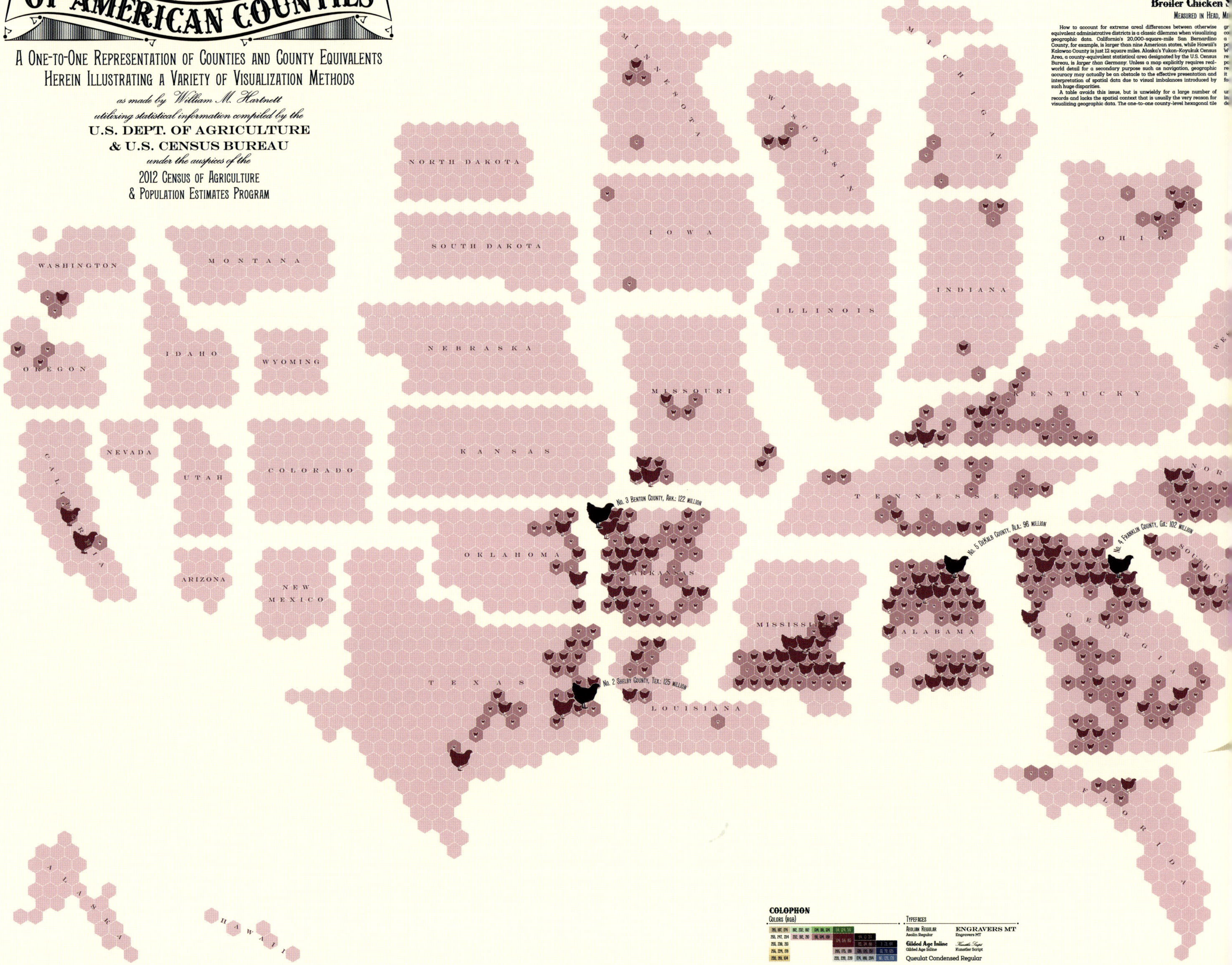

PROPORTIONAL PICTUR

Little Chickens: Charming, Whimsical, F

1 million · 10 million · 25 million · 50 million

Broiler Chicken S

Measured in Head, Mi

How to account for extreme areal differences between otherwise equivalent administrative districts is a classic dilemma when visualizing geographic data. California's 20,000-square-mile San Bernardino County, for example, is larger than nine American states, while Hawaii's Kalawao County is just 12 square miles. Alaska's Yukon-Koyukuk Census Area, a county-equivalent statistical area designated by the U.S. Census Bureau, is *larger than Germany*. Unless a map explicitly requires real-world detail for a secondary purpose such as navigation, geographic accuracy may actually be an obstacle to the effective presentation and interpretation of spatial data due to visual imbalances introduced by such huge disparities.

A table avoids this issue, but is unwieldy for a large number of records and lacks the spatial context that is usually the very reason for visualizing geographic data. The one-to-one county-level hexagonal tile

COLOPHON

Colors (RGB)

215, 197, 174	192, 232, 192	134, 191, 134	54, 124, 54		
255, 247, 234	232, 192, 210	191, 134, 159	124, 54, 85	54, 10, 31	
255, 239, 213				85, 34, 66	3, 26, 64
255, 224, 178			205, 175, 158	128, 120, 150	13, 70, 125
230, 201, 154			220, 220, 220	174, 166, 204	96, 120, 170

Typefaces

Aeolian Regular — Aeolin Regular
ENGRAVERS MT — Engravers MT
Gilded Age Inline — Gilded Age Inline
Kunstler Script — Kunstler Script
Queulat Condensed Regular

. SYMBOLS

HALLENGE

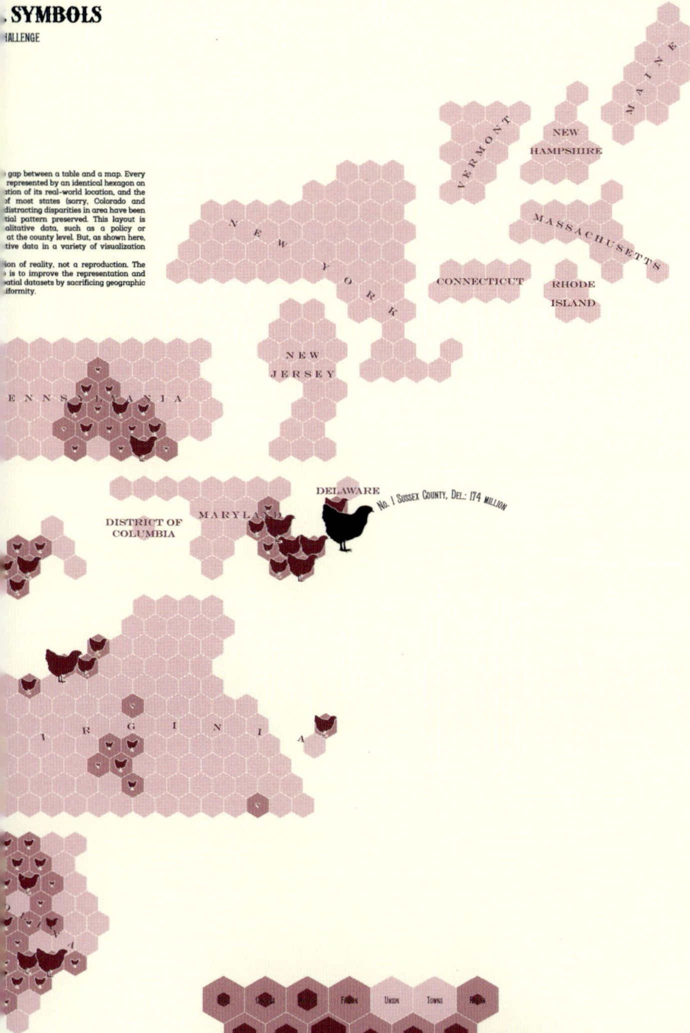

gap between a table and a map. Every represented by an identical hexagon on ation of its real-world location, and the of most states (sorry, Colorado and distracting disparities in area have been tial pattern preserved. This layout is alitative data, such as a policy or at the county level. But, as shown here, tive data in a variety of visualization

ion of reality, not a reproduction. The is to improve the representation and atial datasets by sacrificing geographic iformity.

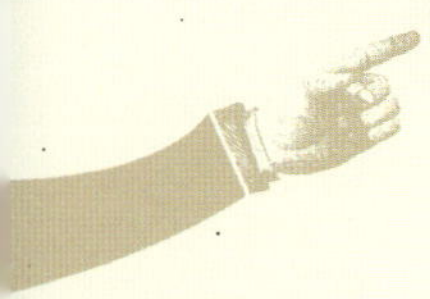

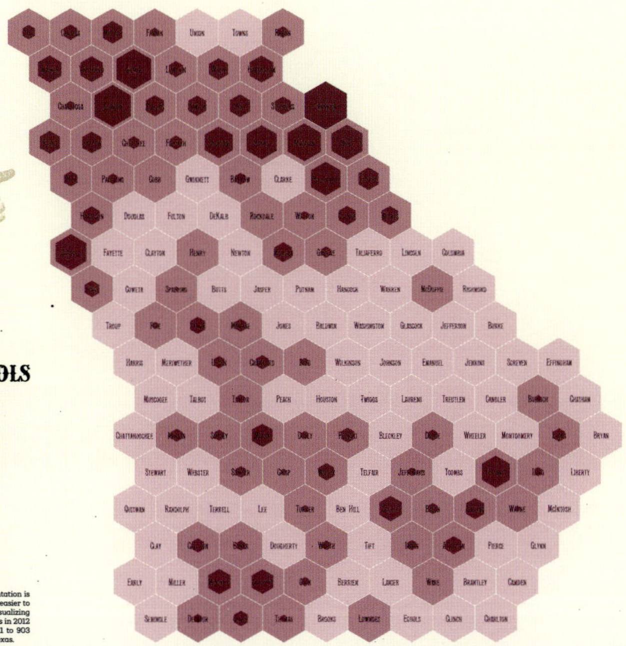

OPORTIONAL SYMBOLS

LESS WHIMSICAL, BETTER PERCEPTUAL ACCURACY

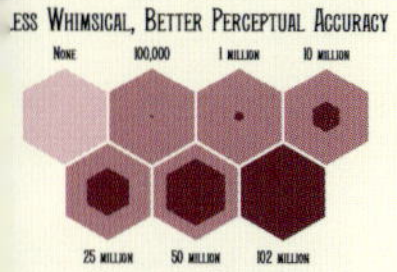

Georgia Broiler Chicken Sales in 2012

MEASURED IN HEAD, NO MINIMUM

traditional example of proportional symbol representation is an using little pictures of chickens, but the scaling is easier to This map also illustrates the tile grid's suitability for visualizing ate. Georgia led the country in sales of broiler chickens in 2012 billion, and its 159 counties, ranging in size from 121 to 903 les, are more numerous than any other state except Texas.

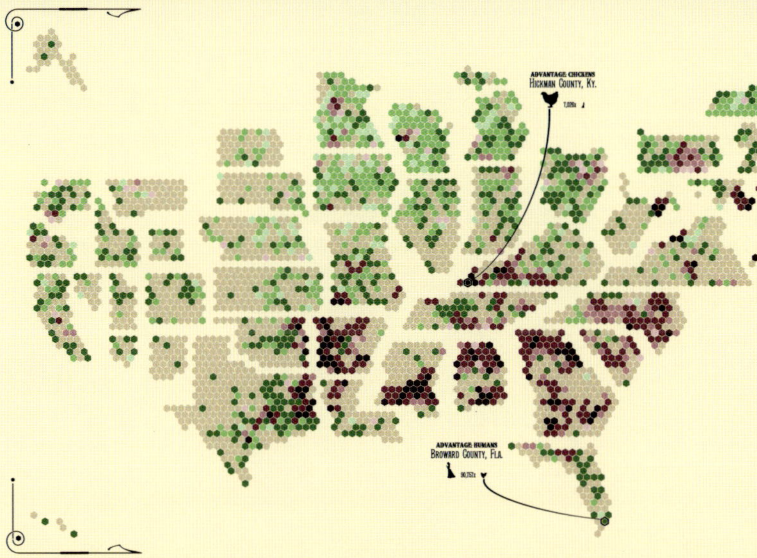

BASIC CHOROPLETH WITH DIVERGING COLORS

FINDING THE FRONTLINES OF AN UNLIKELY CONFLICT

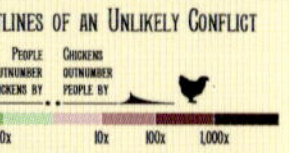

Chickens vs. Humans in 2012

PROPORTIONAL ADVANTAGE OF CHICKENS OR HUMANS

This totally normal, entirely standard-issue choropleth map with a diverging color scheme will probably only be of use in the admittedly unlikely event of nationwide conflict between humans and America's top poultry product. You know, one of those maps. But considering the fact that broiler chickens outnumber us by more than 25-to-1, should we not prepare for any scenario? Green highlights counties in which humans outnumber broiler chickens, while red represents areas where broiler chickens have the edge. Here the tile grid's uniformity negates the disproportionate prominence of extremely large counties, but does not convey any information about the absolute sizes of the human or chicken populations.

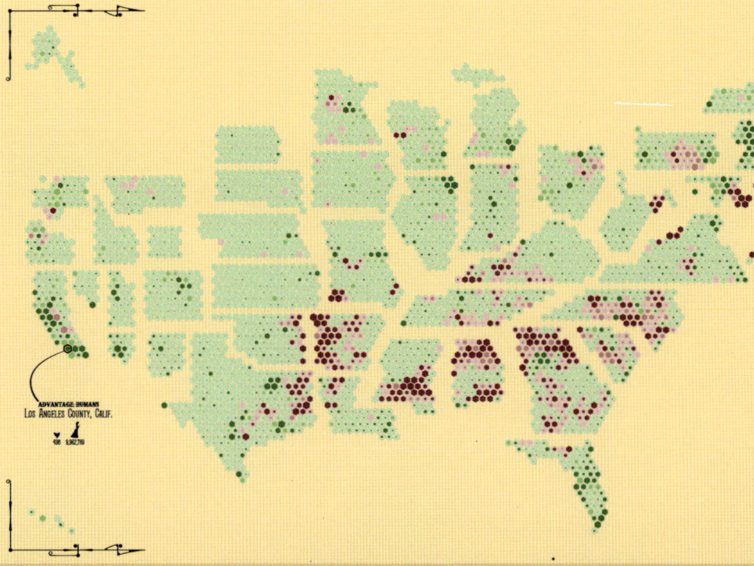

MULTIVARIATE PROPORTIONAL SYMBOLS

ADDING A SECOND VARIABLE TO A SILLY MAP

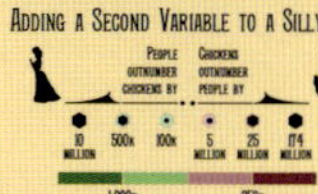

Chickens vs. Humans in 2012

ABSOLUTE AND PROPORTIONAL ADVANTAGE OF CHICKENS OR HUMANS

While the basic choropleth map above does not show the total human or broiler chicken populations, the multivariate proportional symbols here represent both the absolute and proportional advantage of the two sides in this very silly imagined conflict. Note that symbol size and class values are greater for chickens than humans to account for the aforementioned 25-to-1 avian advantage. This is entirely in keeping with the well-established cartographic convention that one human equals 50 broiler chickens. Look it up in any reputable design textbook. The size of hexagonal symbols here represents the numerical advantage of chickens or humans, while the hue and brightness reflect the proportional advantage of each species.

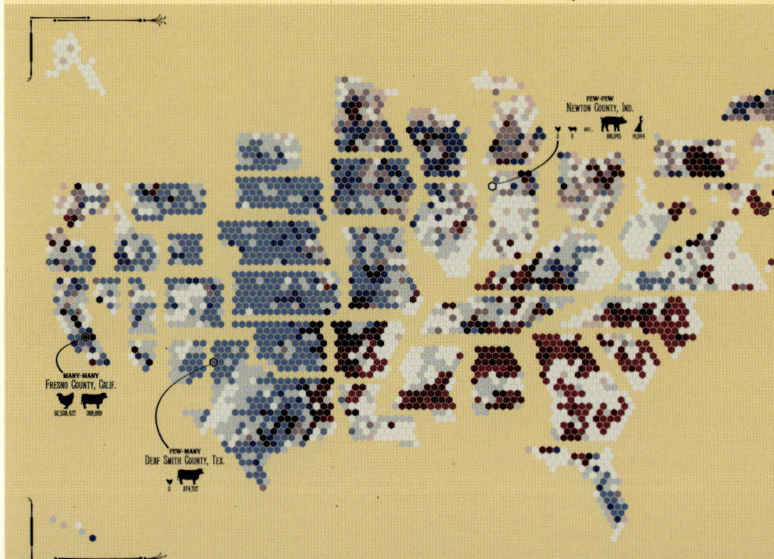

BIVARIATE CHOROPLETH

TOO MANY CLASSES ARE NOT ENOUGH CLASSES

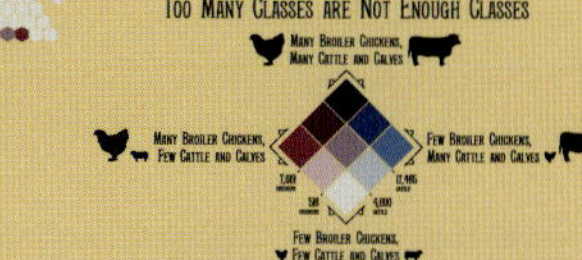

Chickens vs. Cows in 2012

BROILER CHICKEN AND CATTLE AND CALF SALES COMPARED

Reader, you must have known this bivariate choropleth was coming as soon as you saw the words "hexagonal tile grid" 62 inches over there to the left. Best carto combo since fixed-gear bikes and craft beer. As with the univariate choropleth above, the tile grid's uniformity eliminates the visual disparity between geographically large and small counties. Such standardization is particularly useful given the number of categories and colors in this nine-class map. Violet represents counties with many broiler chicken sales but few sales of cattle and calves, and blue highlights counties where the opposite is true. The deep blue mix of chicken violet and cow blue shows counties with many sales of both broilers and cattle.

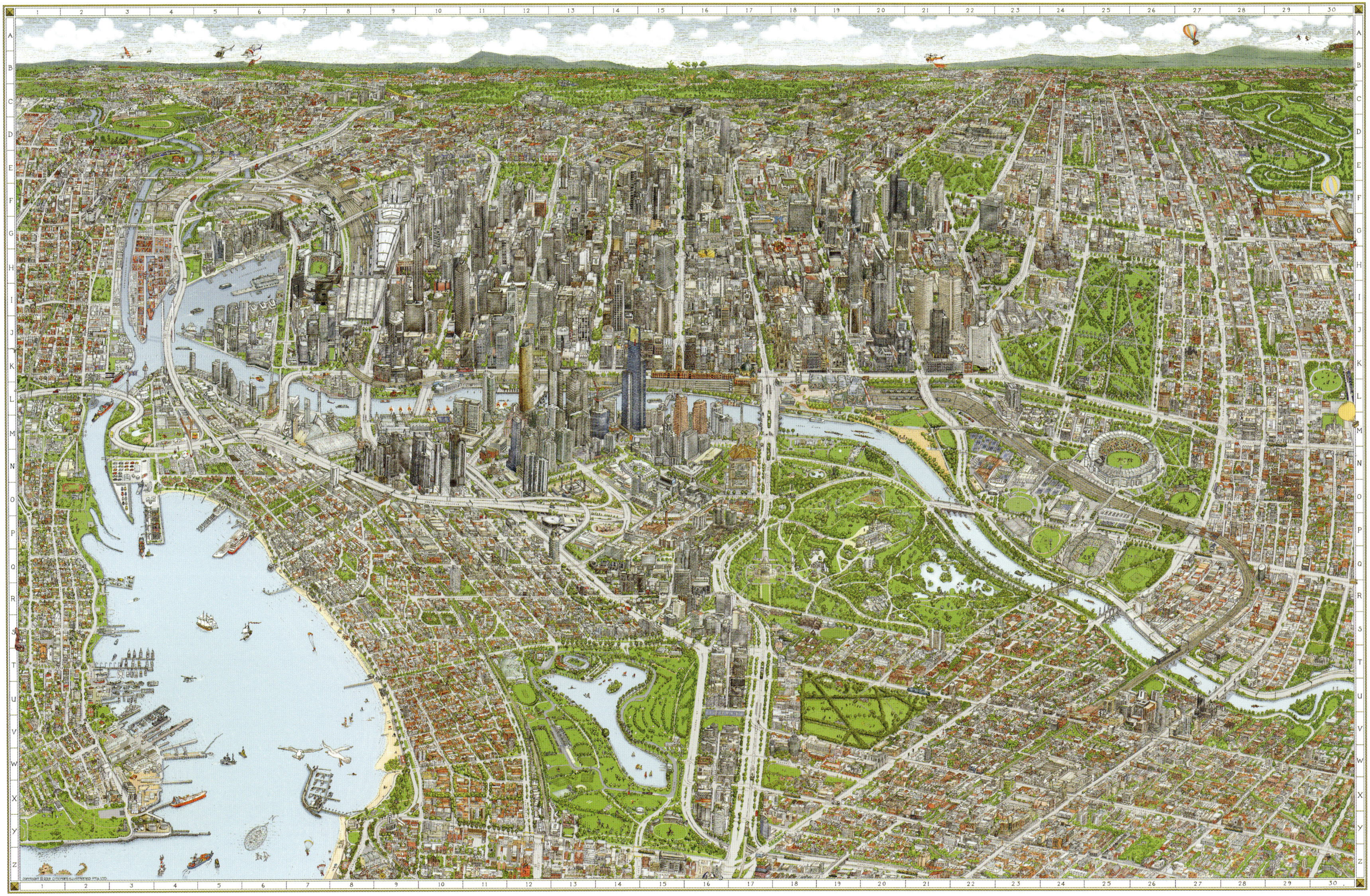

MELBOURNE

VICTORIA • AUSTRALIA

Founder & Producer: Melinda J. Clarke   Chief Illustrator: Lewis W. Brownlie   Character Illustrator: Deborah Young Monk   Digital Colourist: Sean Rodwell

《 The Melbourne Map

Melinda Clarke, Lewis Brownlie, Deborah Young-Monk, and Sean Rodwell — City Maps Illustrated Pty Ltd

The Melbourne Map is an artistic representation of this magnificent city. Hand drawn in intricate detail, it takes the viewer on a journey of discovery. My role in the project is concept, design, research, editing, and publishing. Together, Lewis Brownlie, the chief illustrator, and I spent two and a half years working together to create the line drawing. A large (180 cm × 110 cm) sheet of Arches watercolor paper was attached to some chipboard, and we marked out the boundaries and gridded it into 64 sections. Along the way, we made some hard decisions to completely redraw some areas (redrawing three months' work!) and also to extend the map to the west to include more suburbs. Upon completion, the artwork was scanned and stitched together to create the final map, which also received some contributions from Deborah Young, as character illustrator.

We knew from experience how much it meant to people to have their home, workplace, and suburb accurately reflected, so we wanted to include as much detail as possible. Lewis and I did a ten-hour road trip every couple of weeks, researching the areas that were going to be illustrated next. The rest of the research material was sourced online from publicly available mapping programs. After the whole drawing was penciled in, Lewis began inking-in the intricate detail, using a 0.005 Uni Pin fine-line pen. It was an extremely challenging undertaking, and one that required immense artistic resilience to stick at it, day after day, for two and a half years.

The task of digital coloring was executed with remarkable accuracy by Sean Rodwell, digital colorist, on a 27-inch iMac, using Adobe Photoshop (and a mouse!). A massive undertaking, seeking out color reference online and from our photos—down to the correct color of the one-millimeter illustrated buoys floating on the Yarra River. In August 2019, we released the full-color, giclée limited-edition prints, a poster, and a very popular 1,000-piece jigsaw puzzle. The maps are accompanied by a "Landmarks and Points of Interest Guide" and an entertaining "Can You Find?" list. The black-and-white version was produced as a signed giclée limited-edition art print, plus an offset-press poster.

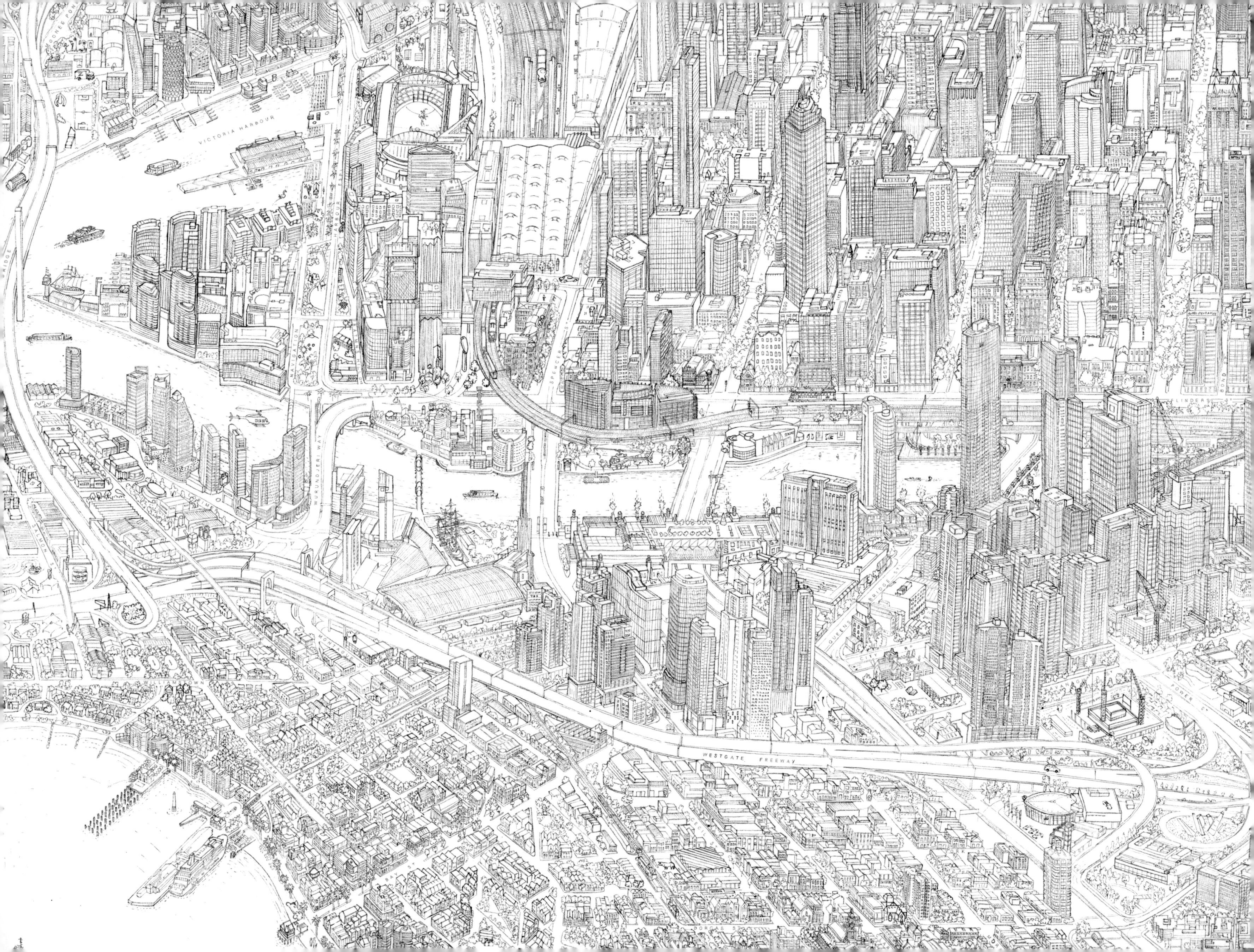

VICTORIA HARBOUR
NAVIGATION DVE
WURUNDJERI WAY
SPENCER ST
FLINDERS ST
BOND ST
QUEENS BRIDGE ST
CITY RD
POWER ST
NORMANBY RD
MONTAGUE ST
MELBOURNE
WESTGATE FREEWAY

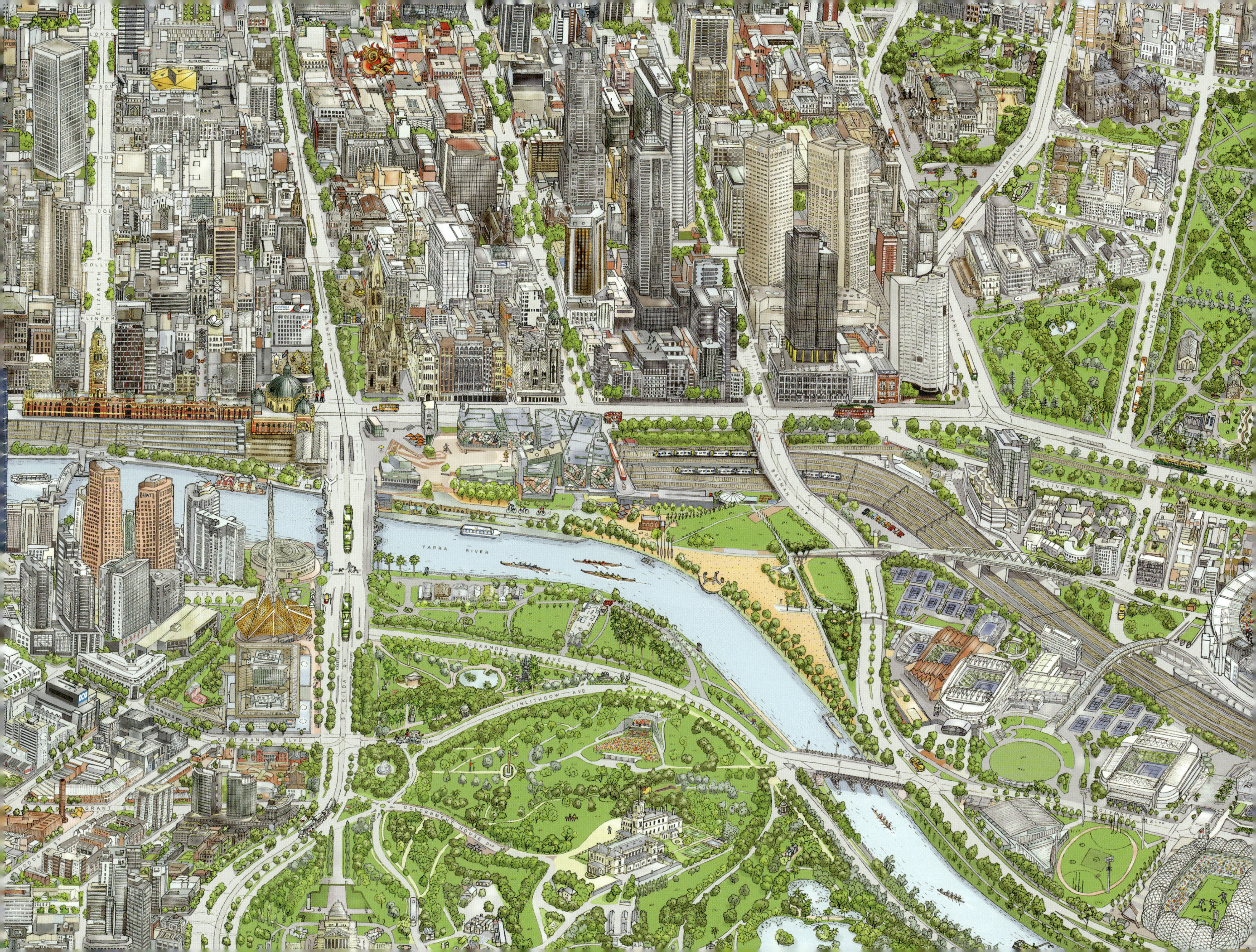
YARRA RIVER
ALEXANDRA AVE
LINLITHGOW AVE
ST KILDA RD

Grave Smoky Mountains National Park

Exploring cemeteries in America's most visited national park

Despite being one of the largest tracts of untamed wilderness in the eastern United States today, Great Smoky Mountains National Park is littered with the remnants of past human settlement.

While earlier western National Parks like Yellowstone and Yosemite were assembled on widely unsettled and government-owned lands, the region that would eventually become Great Smoky Mountains National Park in 1934 had, for the most part, been privately owned. Logging companies, mining towns, and hundreds of small farms and homesteads that totaled over 6,000 individual land parcels were bought out to clear space for the park. Some of these displaced pioneer communities had occupied the region for several generations, while the Cherokee people had lived in the area for more than a thousand years. Life-long Cades Cove resident William Howell Oliver, referring to the churches and cemeteries being handed over to the park, said it felt "like selling our dead". Much of what they left behind was cleared out or crumbled away, though old cabins, churches, and mills still dot the slopes and valleys of the park. Most often, it is the community cemeteries that have survived to serve as a reminder of what came before.

The National Park Service recognizes the locations of more than 150 cemeteries within the park's borders, though many more are thought to exist. History and local lore indicate numerous grave sites lost to nature after nearby settlers vacated the region. Others may have been cleared away intentionally. One story says the headstones of an unlocated cemetery within Cades Cove were carried off by a land owner who wanted to make room for a vegetable garden. Most of these cemeteries that are only known through text and word of mouth are unlikely to ever be located.

The ones we do know of today can be found in all corners of the park, some easily accessible by road and others requiring long, strenuous hikes into the backcountry. Many of the graves are marked by their original headstones, which range from professionally cut and inscribed markers to bare, irregular chunks of stone taken from the nearest riverbed or rocky outcrop. Footstones and complimentary markers placed by later descendants are also common.

The border of this map features inscriptions taken from some of the headstones throughout the park.

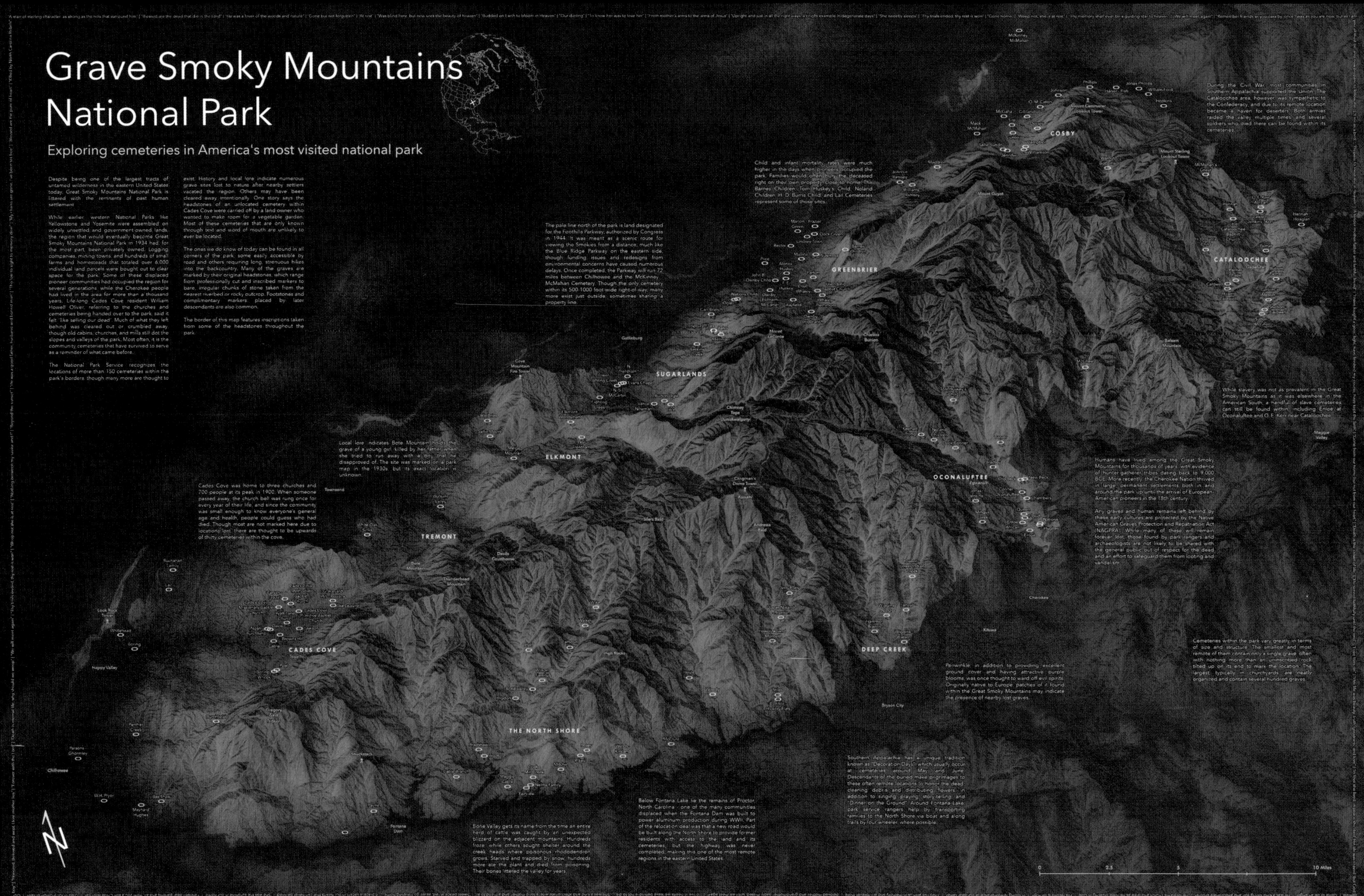

The pale line north of the park is land designated for the Foothills Parkway, authorized by Congress in 1944. It was meant as a scenic route for viewing the Smokies from a distance, much like the Blue Ridge Parkway on the eastern side, though funding issues and redesigns from environmental concerns have caused numerous delays. Once completed, the Parkway will run 72 miles between Chilhowee and the McKinney McMahan Cemetery. Though the only cemetery within its 500-1000 foot-wide right-of-way, many more exist just outside, sometimes sharing a property line.

Child and infant mortality rates were much higher in the days when pioneers occupied the park. Families would often bury the deceased right on their own property, close to home. The Barnes Children, Tom Huskey's Child, Noland Children, H. D. Burris Child, and Lail Cemeteries represent some of those sites.

During the Civil War, most communities in Southern Appalachia supported the Union. The Cataloochee area, however, was sympathetic to the Confederacy, and due to its remote location became a haven for deserters. Both armies raided the valley multiple times, and several soldiers who died there can be found within its cemeteries.

While slavery was not as prevalent in the Great Smoky Mountains as it was elsewhere in the American South, a handful of slave cemeteries can still be found within, including Enloe at Oconaluftee and O. E. Kerr near Cataloochee.

Humans have lived among the Great Smoky Mountains for thousands of years, with evidence of hunter-gatherer tribes dating back to 9,000 BCE. More recently, the Cherokee Nation thrived in large, permanent settlements both in and around the park up until the arrival of European-American pioneers in the 18th century.

Any graves and human remains left behind by these early cultures are protected by the Native American Graves Protection and Repatriation Act (NAGPRA). While many of these will remain forever lost, those found by park rangers and archaeologists are not likely to be shared with the general public out of respect for the dead and an effort to safeguard them from looting and vandalism.

Local lore indicates Bote Mountain holds the grave of a young girl killed by her father when she tried to run away with a boy that he disapproved of. The site was marked on a park map in the 1930s, but its exact location is unknown.

Cades Cove was home to three churches and 700 people at its peak in 1900. When someone passed away, the church bell was rung once for every year of their life, and since the community was small enough to know everyone's general age and health, people could guess who had died. Though most are not marked here due to locations lost, there are thought to be upwards of thirty cemeteries within the cove.

Periwinkle, in addition to providing excellent ground cover and having attractive purple blooms, was once thought to ward off evil spirits. Originally native to Europe, patches of it found within the Great Smoky Mountains may indicate the presence of nearby lost graves.

Cemeteries within the park vary greatly in terms of size and structure. The smallest and most remote of them contain only a single grave, often with nothing more than an uninscribed rock tilted up on its end to mark the location. The largest, typically in churchyards, are neatly organized and contain several hundred graves.

Southern Appalachia has a unique tradition known as "Decoration Days" which usually occur at cemeteries around May and June. Descendants of the buried make pilgrimages to these often remote locations to honor the dead, cleaning debris and distributing flowers, in addition to singing, praying, story-telling and "Dinner on the Ground". Around Fontana Lake, park service rangers help by transporting families to the North Shore via boat and along trails by four wheeler where possible.

Below Fontana Lake lie the remains of Proctor, North Carolina - one of the many communities displaced when the Fontana Dam was built to power aluminum production during WWII. Part of the relocation deal was that a new road would be built along the North Shore to provide former residents with access to the land and its cemeteries, but the highway was never completed, making this one of the most remote regions in the eastern United States.

Bone Valley gets its name from the time an entire herd of cattle was caught by an unexpected blizzard on the adjacent mountains. Hundreds froze, while others sought shelter around the creek heads where poisonous rhododendron grows. Starved and trapped by snow, hundreds more ate the plant and died from poisoning. Their bones littered the valley for years.

《 Grave Smoky Mountains National Park

Aaron Koelker

Most people visit Great Smoky Mountains National Park (GSMNP) to see the bears, waterfalls, and titular misty peaks, but the park has a lot of historical value as well. Unlike many western parks, GSMNP was formed on what were largely privately owned lands, displacing numerous pioneer communities in the process. Most of what they left behind was cleared away or reclaimed by nature, but often the cemeteries remain as a reminder of what came before. I had been reading into this further in the weeks leading up to Halloween and, being interested in local histories and lore—which are both invaluable and under constant threat of being forgotten—was inspired to make this map. Every name on it is an invitation to explore the story of a forgotten community, an isolated family, or a mysterious individual.

Given the season and the subject matter, it was hard not to go for a spooky aesthetic, but I also wanted to convey a sense of mystery and intrigue. I used a subtle mist to obscure the lowlands and a darker overlay to deemphasize areas outside the park. Urban areas were reduced to tiny labels to reinforce an impression of remote wilderness, and muted colors help dictate the tone.

The isometric angle, or parallel view, prevents the cemeteries farthest from the viewer from jumbling together. Their ring shape provides some built-in locational uncertainty. Shadows were added to make them feel more present within the map, giving everything a physical, diorama-esque look, while the Cherokee names scattered throughout remind us that the history of the region did not begin with these European pioneers.

Lastly, I included a text border of epitaphs taken from actual headstones in the park to help the reader make a better connection with the people buried there. I think folks are generally rather detached when looking at maps, since they abstract so much. Some inscriptions were found online, but many I gathered myself following a last-minute, eight-hour drive to the mountains—the last weekend before park campgrounds closed for the year. It can be summed up as: "48 Hours. Hundreds of tombstones. One tropical depression."

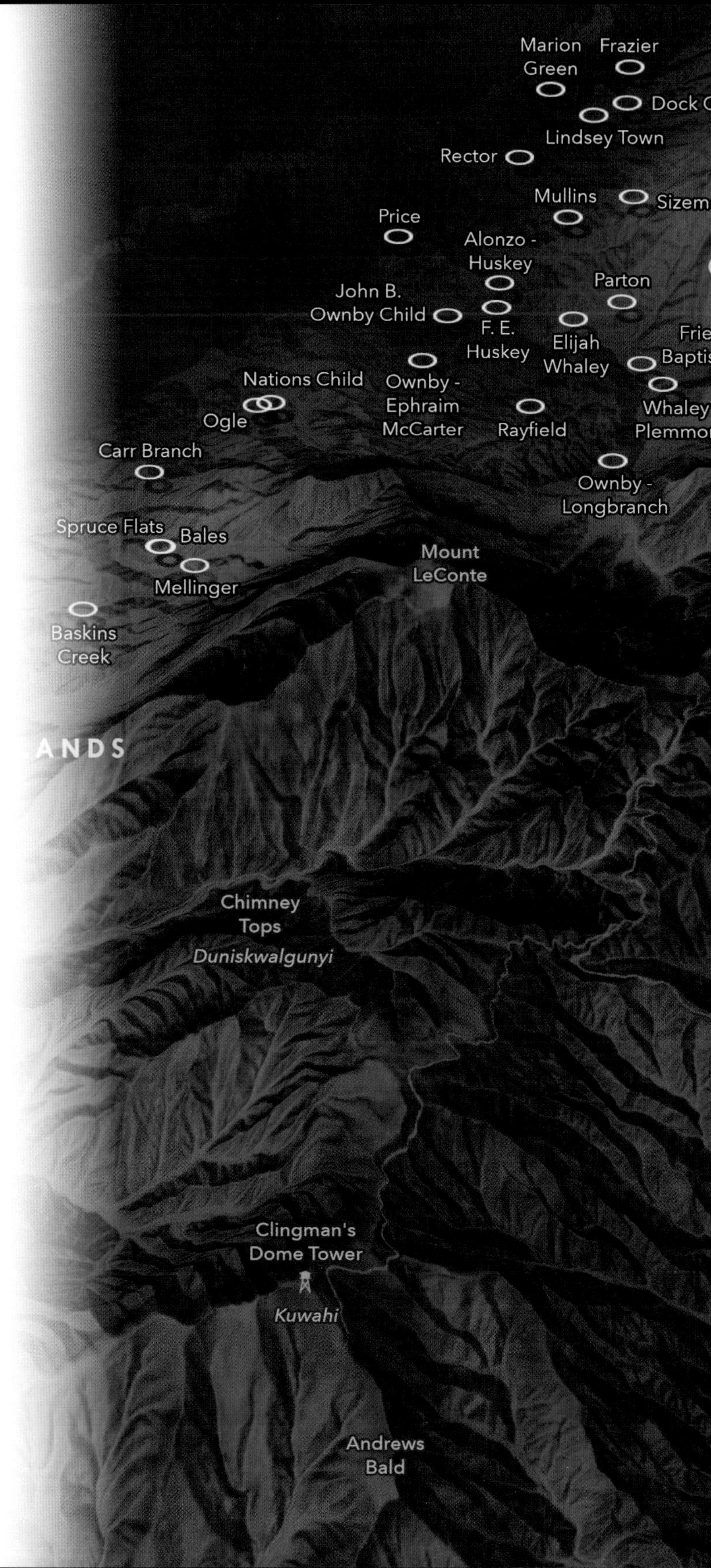

The Teton Fault »

Rick Zeeb — MagnaTerra Cartographic

When a colleague asked if I would create a map documenting a fault study he had recently conducted for the Wyoming State Geological Survey along the Teton fault in Jackson Hole, Wyoming, I jumped at the chance. The iconic beauty of this region would present a great opportunity to create a science-based map that could also appeal to a larger, lay audience. "You mean, faults created these mountains!?" Indeed, they did.

The obvious challenge here is that the fault itself occupies precious little real estate on the map, being literally and cartographically towered over by the adjacent Teton range. And while the faults were to be the focus of the map, I also wanted to show the immensity of the geologic features they've produced, creating a situation where the reader would not only examine the scientific data, but also linger on the entire map, contemplating the beauty and majesty of the region. To this end I used bold colors for faults, while keeping the base map subdued, clear, and precise.

Unfortunately, at this scale fault lines can seem to merge and overlap, becoming a tangle of indistinguishable lines, perhaps even a dreaded "blob." Sometimes I'll give this type of line work a faint drop shadow to pull it slightly off the page. I avoided that here because it tends to change the perception of the actual line location relative to the surface below, and accurate location was key to the study. In the end, I gave each fault a barely perceptible dark outline, just enough so each individual line can be discerned.

Additional techniques I used to give the map a friendly, less scientific feel were extending the hill shade beyond the map boundary, underlying the legend and notes, and placing the map title on a curve (much to the chagrin of the Wyoming State Geologic Survey; thanks for relenting, guys!). I also let the map boundary trail off at the south while closing it off at the north, making it appear thrust up from the bottom of the sheet, not unlike how the Tetons were thrust skyward by the Teton fault.

WYOMING STATE GEOLOGICAL SURVEY
Erin A. Campbell, Director and State Geologist
Laramie, Wyoming
Interpreting the past, providing for the future

OPEN FILE REPORT 2019-1

The Teton Fault

Teton County, Wyoming

by
Mark S. Zellman[1], Christopher B. DuRoss[2], and Glenn D. Thackray[3]
Cartography by Richard B. Zeeb[4]
[1]BGC Engineering, Inc., [2]U.S. Geological Survey, [3]Idaho State University, [4]MagnaTerra Cartographic, Inc.

2019

LOCATION MAP

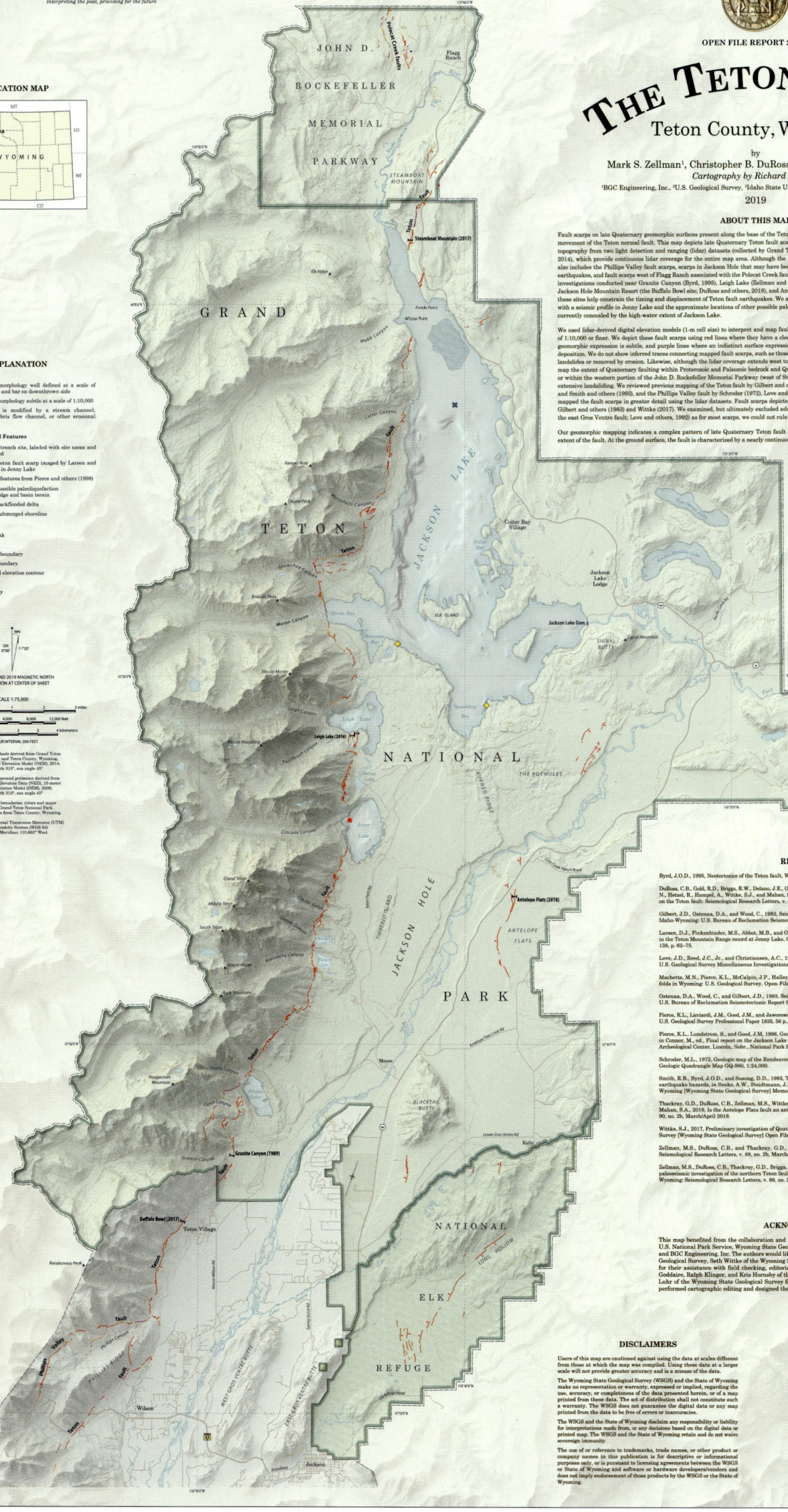

EXPLANATION

Faults

- Fault scarp morphology well defined at a scale of 1:10,000; ball and bar on downthrown side
- Fault scarp morphology subtle at a scale of 1:10,000
- Fault scarp is modified by a stream channel, shoreline, debris flow channel, or other erosional feature

Paleoseismic Sites and Features

- Paleoseismic trench site, labeled with site name and year excavated
- Location of Teton fault scarp imaged by Larsen and others (2016) in Jenny Lake
- Paleoseismic features from Pierce and others (1998)
  - Possible paleoliquefaction ridge and basin terain
  - Backflooded delta
  - Submerged shoreline

Base Data

- Mountain peak
- Airport
- Federal land boundary
- Lidar data boundary
- Lidar-derived elevation contour
- U.S. highway
- State highway
- Minor road

GN 0°00′ MN 11°20′

UTM GRID AND 2019 MAGNETIC NORTH DECLINATION AT CENTER OF SHEET

SCALE 1:75,000

1 0 1 2 3 miles

2,000 0 4,000 8,000 12,000 feet

1 0 1 2 3 4 kilometers

CONTOUR INTERVAL 200 FEET

Base lidar hillshade derived from Grand Teton National Park and Teton County, Wyoming, 1-meter Digital Elevation Model (DEM), 2014; azimuth 315°, sun angle 45°

Base hillshade around perimeter derived from United States Elevation Data (NED), 10-meter Digital Elevation Model (DEM), 2008; azimuth 315°, sun angle 45°

Federal land boundaries, rivers and major lakes from Grand Teton National Park. Road centerlines from Teton County, Wyoming.

Projection: Universal Transverse Mercator (UTM)
World Geodetic System (WGS 84)
Central Meridian: 110.683° West

ABOUT THIS MAP

Fault scarps on late Quaternary geomorphic surfaces present along the base of the Teton Range in Jackson Hole, Wyoming, attest to geologically recent movement of the Teton normal fault. This map depicts late Quaternary Teton fault scarps interpreted from field observations (2015–2018) and surface topography from two light detection and ranging (lidar) datasets (collected by Grand Teton National Park and Teton County Conservation District in 2014), which provide continuous lidar coverage for the entire map area. Although the map emphasizes fault scarps associated with the Teton fault, it also includes the Phillips Valley fault scarps, scarps in Jackson Hole that may have been caused by faulting or strong ground shaking from Teton fault earthquakes, and fault scarps west of Flagg Ranch associated with the Polecat Creek faults. The map shows the location of five paleoseismic (fault-trench) investigations conducted near Granite Canyon (Byrd, 1995), Leigh Lake (Zellman and others, 2017), Steamboat Mountain (Zellman and others, 2018), Jackson Hole Mountain Resort (the Buffalo Bowl site; DuRoss and others, 2018), and Antelope Flats (Thackray and others, 2019). Paleoseismic data from these sites help constrain the timing and displacement of Teton fault earthquakes. We also show where Larsen and others (2016) imaged the Teton fault with a seismic profile in Jenny Lake and the approximate locations of other possible paleoseismic features described by Pierce and others (1998) that are currently concealed by the high-water extent of Jackson Lake.

We used lidar-derived digital elevation models (1-m cell size) to interpret and map fault scarps of likely late Quaternary age and discernible at a scale of 1:10,000 or finer. We depict these fault scarps using red lines where they have a clearly recognizable geomorphic expression, orange lines where the geomorphic expression is subtle, and purple lines where an indistinct surface expression results from surface modification from stream erosion and/or deposition. We do not show inferred traces connecting mapped fault scarps, such as those in areas where fault scarps may be concealed beneath lakes and landslides or removed by erosion. Likewise, although the lidar coverage extends west to the Wyoming-Idaho border, we did not attempt to interpret and map the extent of Quaternary faulting within Proterozoic and Paleozoic bedrock and Quaternary sediments west of the Teton and Phillips Valley faults or within the western portion of the John D. Rockefeller Memorial Parkway (west of Steamboat Mountain) where surface topography is complicated by extensive landsliding. We reviewed previous mapping of the Teton fault by Gilbert and others (1983), Love and others (1992), Ostenaa and others (1993), and Smith and others (1993), and the Phillips Valley fault by Schroder (1972), Love and others (1992), and Machette and others (2001), but we generally mapped the fault scarps in greater detail using the lidar datasets. Fault scarps depicted in eastern and southeastern Jackson Hole were modified from Gilbert and others (1983) and Wittke (2017). We examined, but ultimately excluded additional faults previously mapped throughout Jackson Hole (e.g., the east Gros Ventre fault; Love and others, 1992) as for most scarps, we could not rule out a glacial or fluvial origin.

Our geomorphic mapping indicates a complex pattern of late Quaternary Teton fault scarps that continue north and south of the previously mapped extent of the fault. At the ground surface, the fault is characterized by a nearly continuous alignment of fault scarps that vertically offset late Pleistocene glacial, and late Pleistocene to Holocene alluvial-fan and landslide deposits. These scarps indicate that the surface trace of the Teton fault is at least 72 km long, from ~3 km southwest of Wilson to ~6 km north of Jackson Lake, near Steamboat Mountain. Locally, the fault trace is complex and includes fault bends, stepovers, and subparallel zones of synthetic and antithetic faulting. At a regional scale, fault scarps are obscured by a landslide north of Colter Canyon (Ostenaa and others, 1993) and an apparent landslide north of Teton Village, and submerged beneath Jenny Lake and Jackson Lake. Bends in the fault trace occur at the mouths of Waterfalls Canyon and Snowshoe Canyon, Leigh Lake, and between Phelps and Taggart Lakes. We mapped apparent fault stepovers near Phelps Lake and Leigh Lake.

Our mapping includes groups of small (less than ~2 m high), generally north-trending scarps on late Quaternary surfaces in central Jackson Hole, between the Potholes and the National Elk Refuge, and north of Jackson Lake near Flagg Ranch. We interpreted a tectonic origin for these scarps as they continue across and vertically displace various Quaternary geomorphic surfaces (~mid-Pleistocene in age and younger; Pierce and others, 2018). Except for the scarps at the Potholes, the scarps in central Jackson Hole mostly face west, opposite (antithetic to) the Teton fault. Scarps mapped west of Flagg Ranch are likely the southern extent of the Polecat Creek faults, a subset of the Snake River Caldera faults (Ostenaa and others, 1993; Machette and others, 2001) which continue to the north beyond the extent of the lidar coverage. These scarps taper southward and become indistinct in the lidar on the north side of the Snake River.

REFERENCES

Byrd, J.O.D., 1995, Neotectonics of the Teton fault, Wyoming: Salt Lake City, University of Utah, PhD. dissertation, 213 p.

DuRoss, C.B., Gold, R.D., Briggs, R.W., Delano, J.E., Ostenaa, D.A., Zellman, M.S., Scharer, K.R., McPhillips, D., Cholewinski, N., Hetzel, R., Hampel, A., Wittke, S.J., and Mahan, S.A., 2018, Preliminary paleoseismology of the Buffalo Bowl trench site on the Teton fault: Seismological Research Letters, v. 89, no. 2b, March/April 2018.

Gilbert, J.D., Ostenaa, D.A., and Wood, C., 1983, Seismotectonic study Jackson Lake Dam and Reservoir-Minidoka project, Idaho-Wyoming: U.S. Bureau of Reclamation Seismotectonic Report 83-8, 122 p.

Larsen, D.J., Finkenbinder, M.S., Abbot, M.B., and Ofstun, A.R., 2016, Deglaciation and postglacial environmental changes in the Teton Mountain Range record at Jenny Lake, Grand Teton National Park, Wyoming: Quaternary Science Reviews, v. 138, p. 62–75.

Love, J.D., Reed, J.C., Jr., and Christiansen, A.C., 1992, Geologic map of Grand Teton National Park, Teton County, WY: U.S. Geological Survey Miscellaneous Investigations Series, Map I-2031.

Machette, M.N., Pierce, K.L., McCalpin, J.P., Haller, K.M., and Dart, R.L., 2001, Map and data for Quaternary faults and folds in Wyoming: U.S. Geological Survey, Open-File Report 01-461, 154 p., 1 pl., scale 1:750,000.

Ostenaa, D.A., Wood, C., and Gilbert, J.D., 1993, Seismotectonic study for Grassy Lake Dam-Minidoka project, Wyoming: U.S. Bureau of Reclamation Seismotectonic Report 93-3, 68 p., 1 pl.

Pierce, K.L., Licciardi, J.M., Good, J.M., and Jaworowski, C., 2018, Pleistocene glaciation of the Jackson Hole area, Wyoming: U.S. Geological Survey Professional Paper 1835, 56 p., available at https://doi.org/10.3133/pp1835.

Pierce, K.L., Lundstrom, S., and Good, J.M, 1998, Geologic setting of archeological sites in the Jackson Lake area, Wyoming, *in* Connor, M., ed., Final report on the Jackson Lake archeological project, Grand Teton National Park, Wyoming: Midwest Archeological Center, Lincoln, Nebr., National Park Service Technical Report No. 46, p. 29–48 and 19 figures on p. 222-242.

Schroder, M.L., 1972, Geologic map of the Rendezvous Peak quadrangle, Teton County, Wyoming: U.S. Geological Survey Geologic Quadrangle Map GQ-980, 1:24,000.

Smith, R.B., Byrd, J.O.D., and Susong, D.D., 1993, The Teton fault, Wyoming—Seismotectonics, Quaternary history, and earthquake hazards, *in* Snoke, A.W., Steidtmann, J.R., and Roberts, S.M., eds., Geology of Wyoming: Geological Survey of Wyoming [Wyoming State Geological Survey] Memoir 5, p. 628–667.

Thackray, G.D., DuRoss, C.B., Zellman, M.S., Wittke, S.J., Gold, R.D., Delano, J.E., Jobe, J.A.T., Hille, M., Grasso, K., and Mahan, S.A., 2019, Is the Antelope Flats fault an antithetic rupture of the Teton fault?: Seismological Research Letters, v. 90, no. 2b, March/April 2019.

Wittke, S.J., 2017, Preliminary investigation of Quaternary faults in eastern Jackson Hole, Wyoming: Wyoming Geological Survey [Wyoming State Geological Survey] Open File Report 2017-1, March 2017.

Zellman, M.S., DuRoss, C.B., and Thackray, G.D., 2017, Paleoseismic investigation of the Teton fault at Leigh Lake: Seismological Research Letters, v. 88, no. 2b, March/April 2017.

Zellman, M.S., DuRoss, C.B., Thackray, G.D., Briggs, R.W., Cholewinski, N., Reyes, T., Patton, N., and Mahan, S.A., 2018, A paleoseismic investigation of the northern Teton fault at the Steamboat Mountain trench site, Grand Teton National Park, Wyoming: Seismological Research Letters, v. 89, no. 2b, March/April 2018.

ACKNOWLEDGMENTS

This map benefited from the collaboration and input from many individuals and organizations. We thank the U.S. National Park Service, Wyoming State Geological Survey, U.S. Geological Survey, Idaho State University, and BGC Engineering, Inc. The authors would like to specifically thank Rich Briggs and Jaime Delano of the U.S. Geological Survey, Seth Wittke of the Wyoming State Geological Survey, and Dean Ostenaa of Ostenaa Geologic for their assistance with field checking, editorial review, and comments. We would also like to thank Jeanne Goddaire, Ralph Klinger, and Kris Hornsby of the U.S. Bureau of Reclamation for their comments and Suzanne Luhr of the Wyoming State Geological Survey for her editorial review. Rick Zeeb of MagnaTerra Cartographic performed cartographic editing and designed the graphical layout of this map.

DISCLAIMERS

NOTICE TO USERS OF INFORMATION FROM THE WYOMING STATE GEOLOGICAL SURVEY

Preferred Citation: Zellman, M.S., DuRoss, C.B., and Thackray, G.D., 2019, The Teton fault: Wyoming State Geological Survey Open File Report 2019-1, scale 1:75,000.

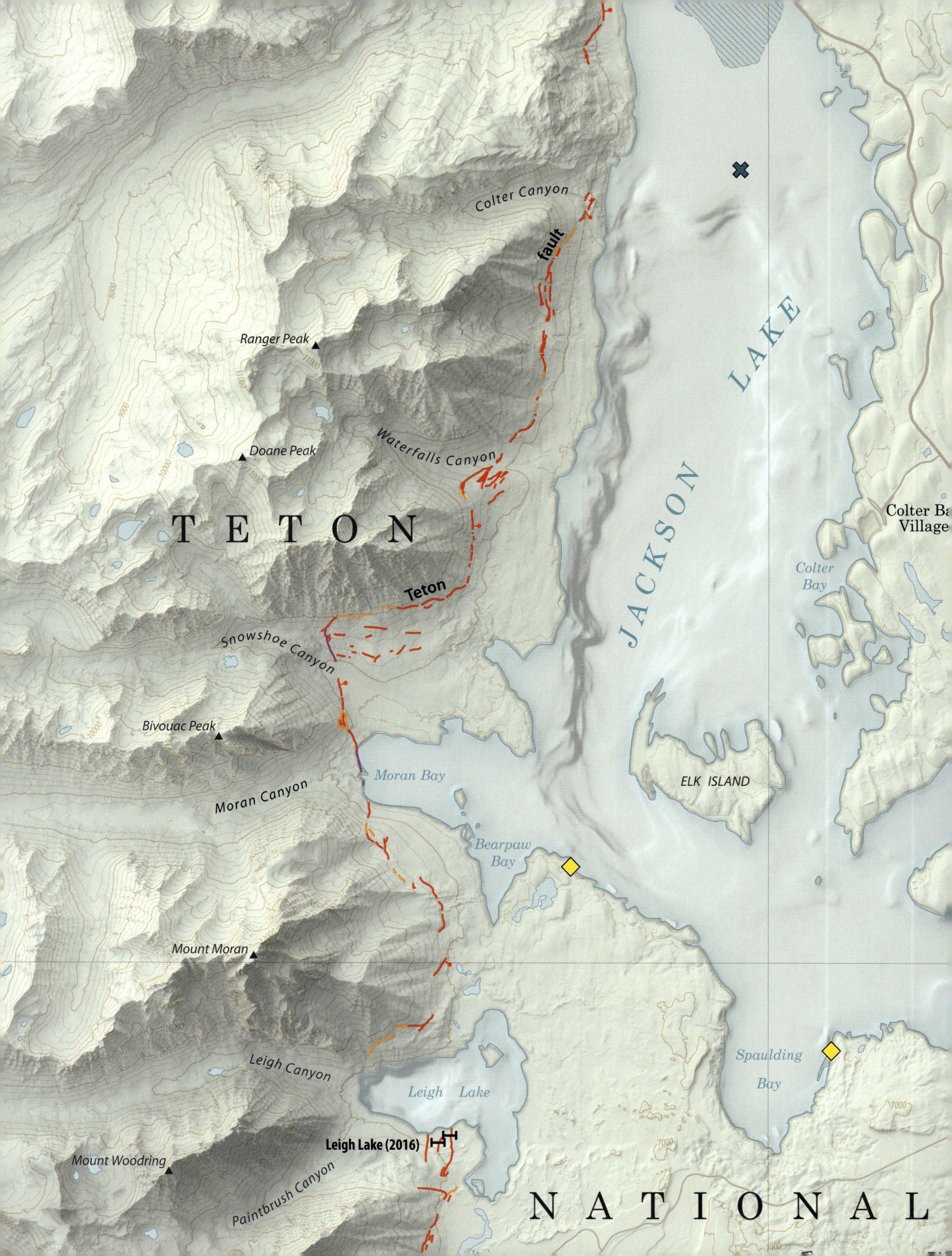
Colter Canyon
fault
Ranger Peak
Doane Peak
Waterfalls Canyon
TETON
Teton
Snowshoe Canyon
Bivouac Peak
Moran Canyon
Moran Bay
Bearpaw Bay
Mount Moran
Leigh Canyon
Leigh Lake
Leigh Lake (2016)
Mount Woodring
Paintbrush Canyon
NATIONAL
JACKSON LAKE
ELK ISLAND
Colter Bay
Colter Bay Village
Spaulding Bay

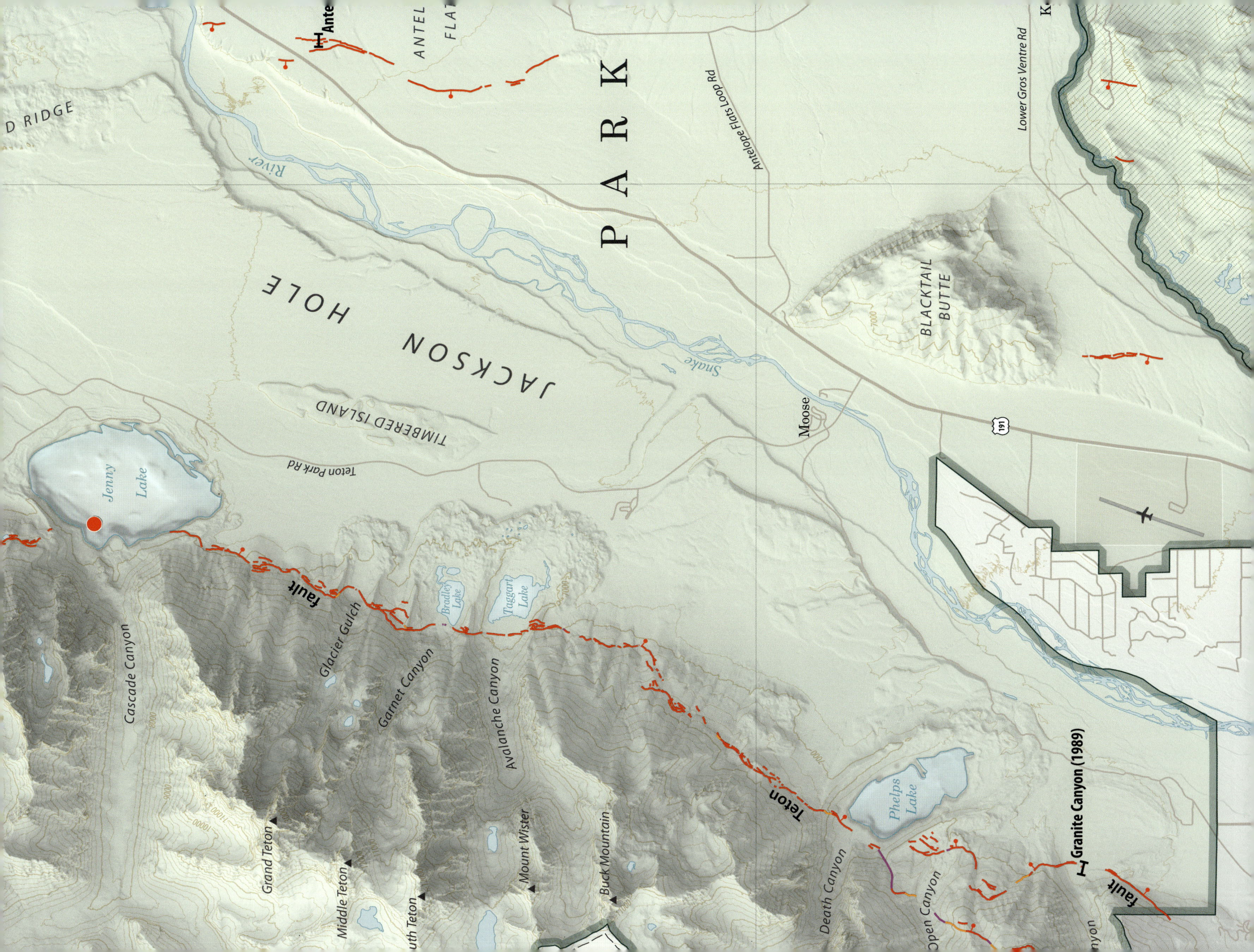
PARK
JACKSON HOLE
TIMBERED ISLAND
Teton Park Rd
Antelope Flats Loop Rd
Lower Gros Ventre Rd
BLACKTAIL BUTTE
Snake River
Moose
191
D RIDGE
ANTEL
FLA
Jenny Lake
Bradley Lake
Taggart Lake
Phelps Lake
Cascade Canyon
Glacier Gulch
Garnet Canyon
Avalanche Canyon
Death Canyon
Open Canyon
Grand Teton
Middle Teton
Mount Wister
Buck Mountain
Teton
fault
Granite Canyon (1989)
Fault

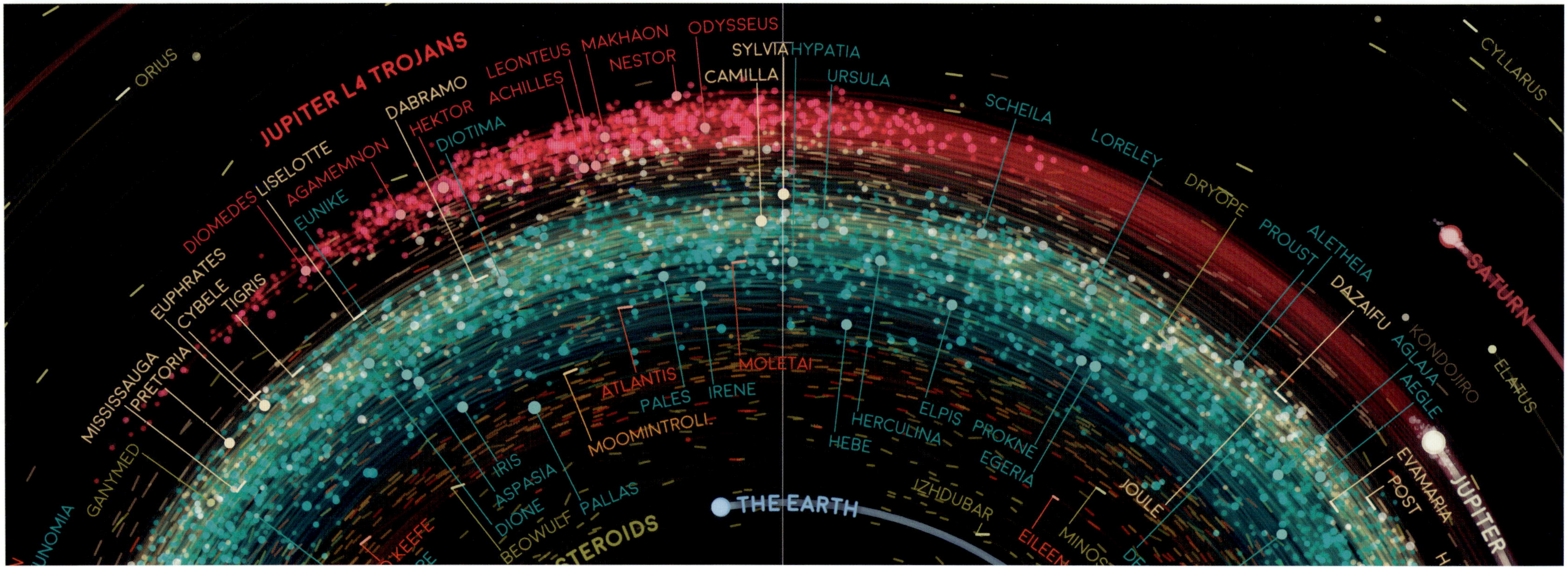

An Orbit Map of the Solar System »

Eleanor Lutz

I really enjoyed learning more about our solar system through this project. I wanted to learn more about data design through creating this map, and to enjoy exploring a type of geography I wasn't very familiar with. I was also surprised by the sheer number of asteroids scientists have mapped, especially smaller ones.

This map visualizes the orbits of more than 18,000 asteroids in the solar system, as well as those of the planets and their moons. It includes everything humans have discovered that's more than 10 km in diameter—about 10,000 asteroids—as well as 8,000 random objects of unknown size. The colors in the map were one of my favorite things to design. Each type of asteroid is visualized in a different color, and I enjoyed picking neon, retro-inspired colors to contrast with the dark background.

One of the most challenging parts of this map was figuring out how to make everything legible. When I started looking into the orbit pattern data provided by NASA, I realized that things in the solar system are distributed somewhat logarithmically. There are exponentially more things closer to the sun, and exponentially more smaller things than big things. So I used a radial logarithmic scale—which is a way of showing a very wide range of numbers on the same visualization—to include large things, faraway things, and tiny, nearby things all on the same map. A logarithmic scale isn't often used for mapping, because it distorts the distance between objects. But in this case, I really liked the effect of showing the entire solar system, but still keeping individual asteroids distinguishable.

This map was built entirely using open data from NASA and the Jet Propulsion Laboratory.

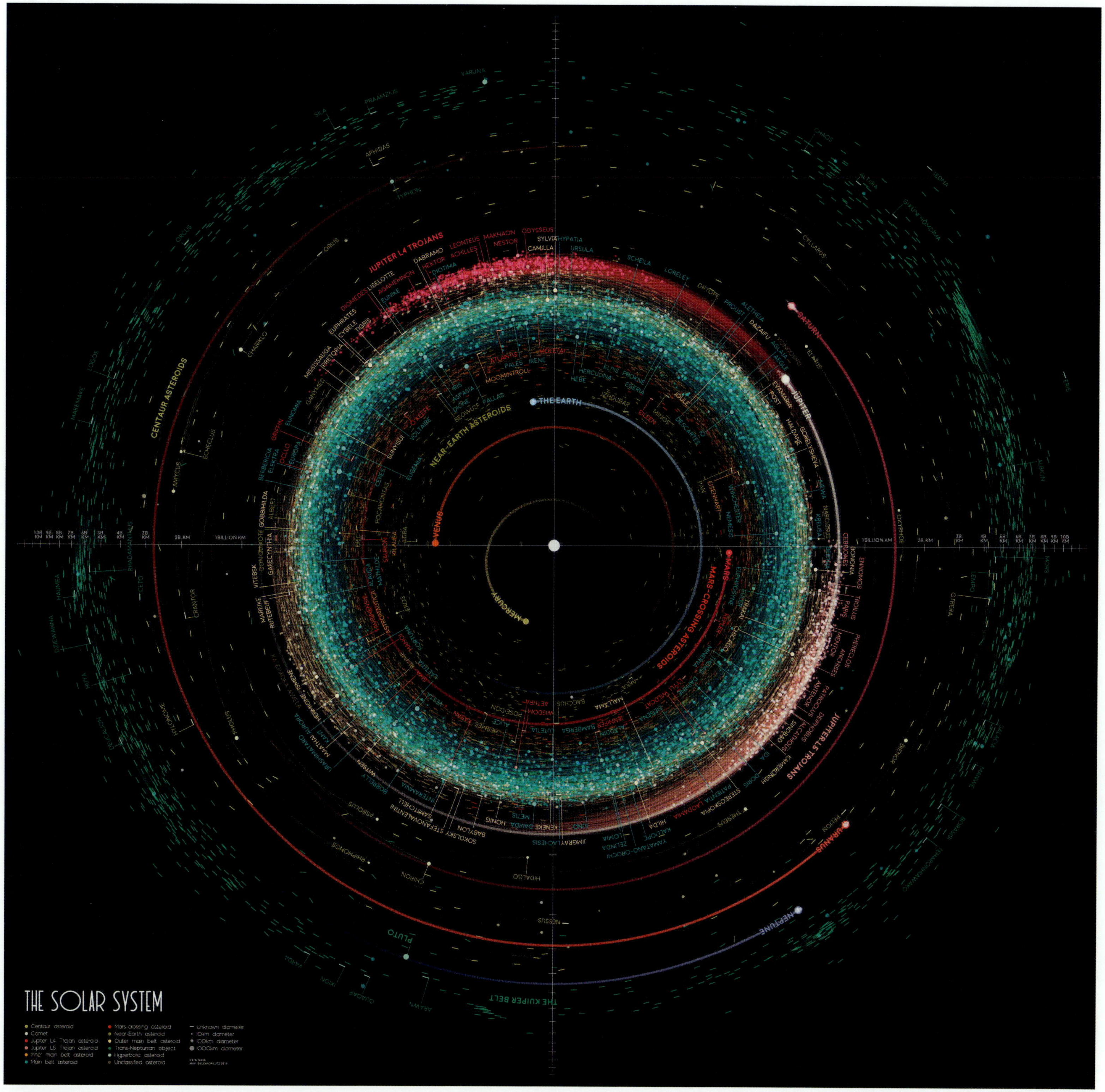

THE SOLAR SYSTEM
Centaur asteroid
Comet
Jupiter L4 Trojan asteroid
Jupiter L5 Trojan asteroid
Inner main belt asteroid
Main belt asteroid
Mars-crossing asteroid
Near-Earth asteroid
Outer main belt asteroid
Trans-Neptunian object
Hyperbolic asteroid
Unclassified asteroid
Unknown diameter
10km diameter
100km diameter
1000km diameter
DATA: NASA
JUPITER L4 TROJANS
JUPITER L5 TROJANS
CENTAUR ASTEROIDS
NEAR-EARTH ASTEROIDS
MARS-CROSSING ASTEROIDS
THE KUIPER BELT
THE EARTH
VENUS
MERCURY
MARS
JUPITER
SATURN
URANUS
NEPTUNE
PLUTO
1BILLION KM
2B KM

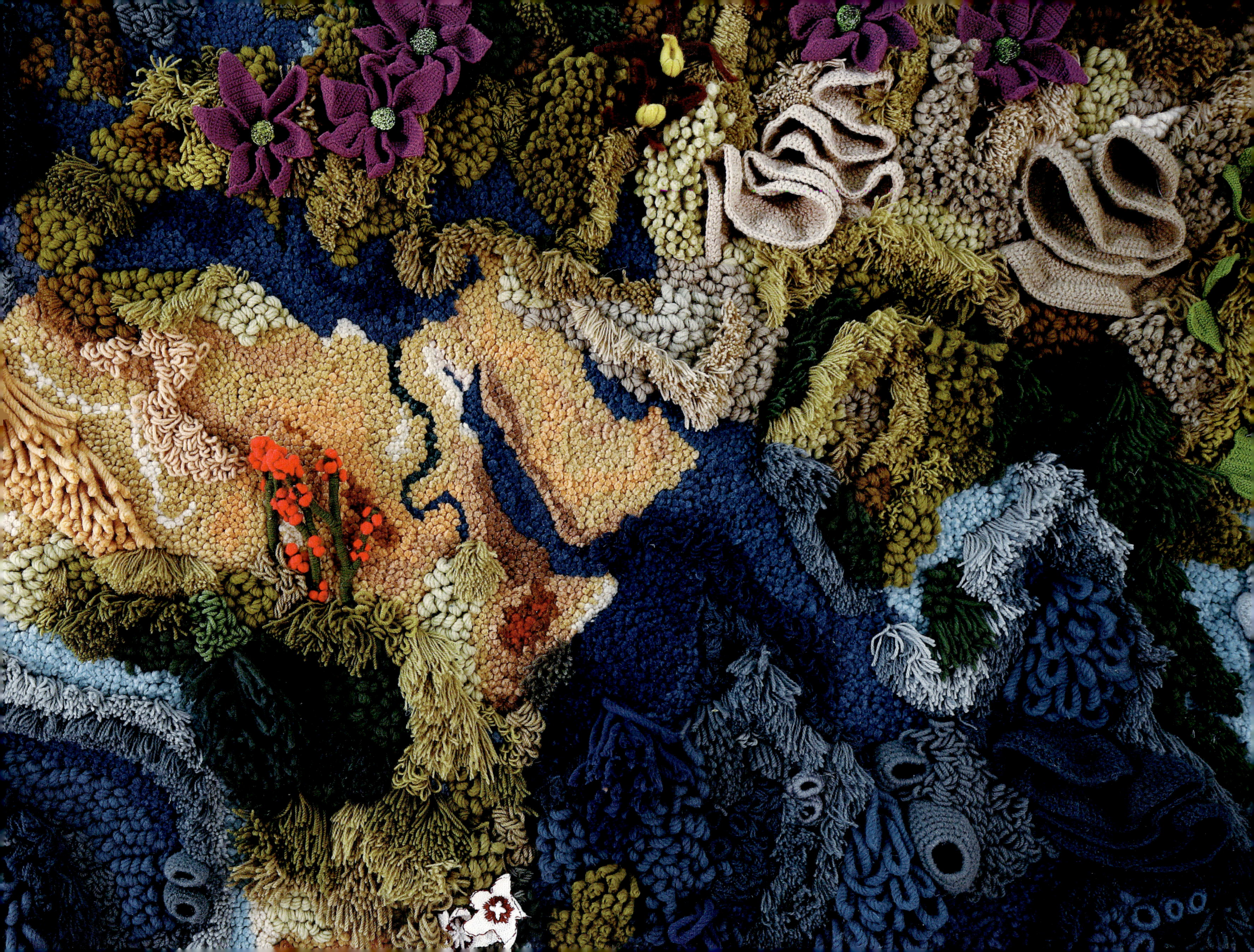

《 Botanical Worldmap Tapestry

Vanessa Barragão, in partnership with Kew Gardens and Heathrow Airport

Botanical Tapestry is an artwork I developed and created, along with the rest of Studio Vanessa Barragão, to celebrate the partnership between Heathrow Airport and Kew Gardens. This tapestry, measuring 600 × 200 cm, is installed in Heathrow Airport Terminal 2, in the departure forecourt. It was important that the piece be large in order to catch the attention of those passing by in such a busy area. Larger size allowed for more detail, such as more islands.

The work is 100 percent handmade, using ancestral techniques and recycled materials. It took more than 520 hours to weave. A total of 8 kg of jute and cotton, and 42 kg of recycled wool were used. This artwork is a vibrant, gestural abstraction of the world map. With the use of latch hook, felt needle, crochet hook, and carving tools and techniques, each continent is rendered in colors and textures reflective of its unique geographic characteristics.

The artwork highlights present environmental concerns, as the piece is punctuated with such threatened species as those of the coral reefs, and protected plants around the world—these include *Cypripedium calceolus* (lady's slipper orchid) and *Pulsatilla vulgaris* (pasque flower) in Europe, *Ginkgo biloba* in China, and the orchid *Bulbophyllum ankylochele* in Indonesia.

Through the continuous effort to make use of upcycled textiles, we demonstrated a refusal to contribute to the production of industrial waste and energy use. Our materials were entirely reclaimed from the deadstock of local factories, which were then cleaned and utilized for the handcrafted, artisanal work. The effort is a response to the dyeing processes, energy consumption by production machinery, and disposal methods of the industry, recognizing the harmful effects of these processes, particularly on the oceans, which absorb 90 percent of the excess heat trapped in the Earth's atmosphere.

Overall, the *Botanical Tapestry* serves as a reminder of the environmental implications across a global scale together with the great diversity of the natural world. With all the pollution that the humans produce, the different ecosystems are suffering and changing, proven by the number of species struggling to survive.

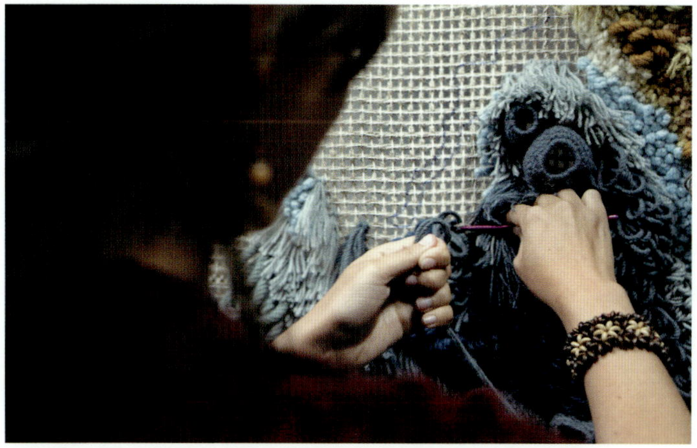

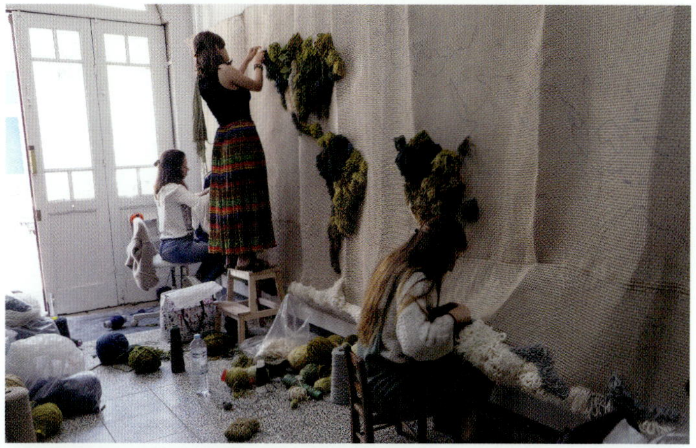

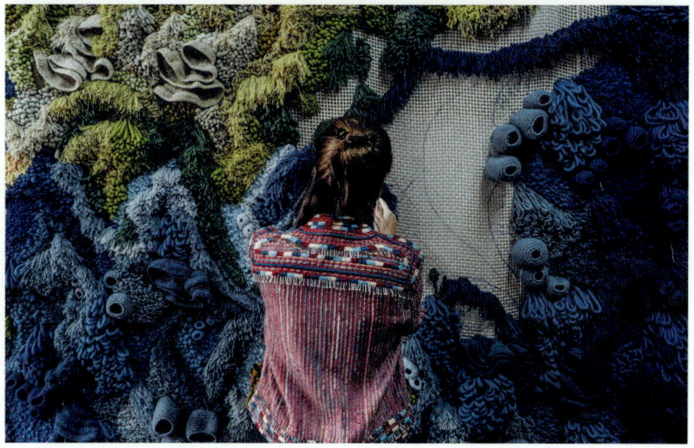

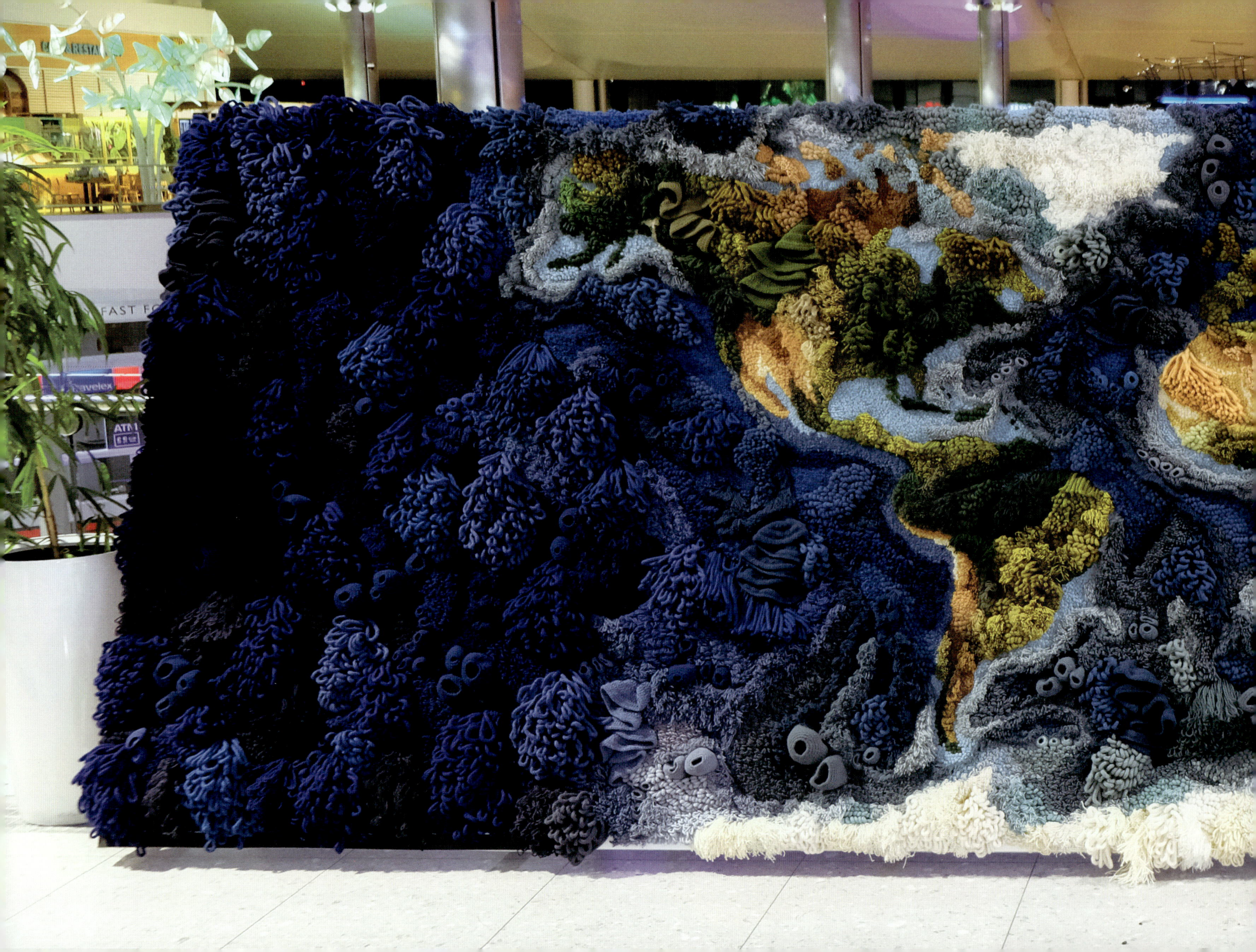
ATM

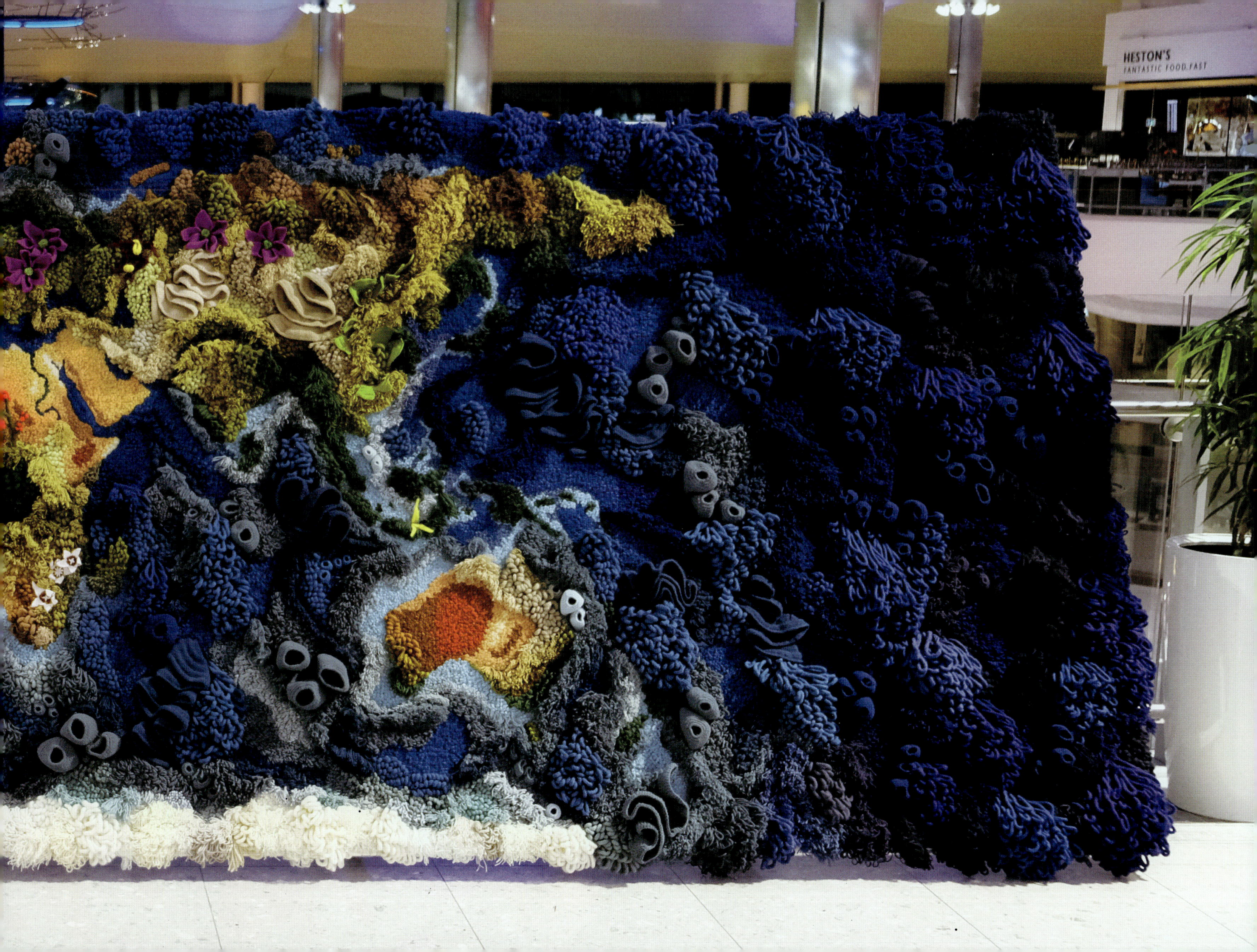
HESTON'S
FANTASTIC FOOD.FAST

atlasofdesign.org